BELGIUM AND POLAND IN INTERNATIONAL RELATIONS 1830—1831

BELGIUM AND POLAND IN INTERNATIONAL RELATIONS 1830-1831

BY

J. A. BETLEY, Ph.D.

MOUTON & Co. · 1960 · 's-GRAVENHAGE

Proefschrift Nijmegen 1960

PRINTED IN THE NETHERLANDS

ACKNOWLEDGMENTS

A thesis on the present subject was first presented to the University of Bristol for the degree of M.A. and the writer therefore wishes to acknowledge the University's permission to use the material then collected (and since substantially enlarged) for the purpose of writing the present work.

I am very much obliged to the late Countess Mountbatten of Burma for permission to use the Broadlands Archives of Lord Palmerston and to publish some documents which appear as appendices to this book. Mrs. Blois, the Curator of the Broadlands Collections, allowed me to see the papers at her home and very kindly copied several documents which were necessary for this work. Mr. and Mrs. R. Bagot, of Levens Hall, Westmorland, kindly offered their hospitality to me when I studied the papers of Sir Charles Bagot and they gave me permission to publish them.

The staff of various institutions assisted me with my research and I especially wish to thank Dr. C. Chowaniec, Curator of the Bibliothèque Polonaise in Paris, Messrs. Avelingh and Ribberink of the Algemeen Rijksarchief at The Hague and Mr. A. Taylor Milne, Secretary and Librarian at the University of London Institute of Historical Research.

I am very grateful to my teachers, Miss E. E. Butcher of the University of Bristol, and Professor L.J. Rogier, of the University of Nijmegen. It is difficult to mention all the other teachers, friends and colleagues who in one way and another helped me during my studies and I am aware that I may be making an imperfect selection when naming only Mrs. E. Arens, Dr. S. Bóbr-Tylingo, General M. Kukiel, Dr. R. F. Leslie, Dr. P. Skwarczyński, Dr. F. E. M. Vercauteren and the late Professor A. Żółtowski.

Lt. Col. L. S. Forrer, Fellow of the Royal Numismatic Society, corrected the bulk of the English text, while Mrs. E. Arens and Mr. D. C. Harris, B. A. (Cantab) made additional corrections. I am indebted to Fr. A. H. Hollaardt, O. P., for checking French quotations and texts.

My full time studies were made possible by a grant from the former Committee for the Education of Poles in Great Britain and by the Curators of the University of Nijmegen. I also obtained financial assistance from the Nuyensfonds and from the Thijmgenootschap. The Netherlands Organisation for the Advancement of Pure Research is sponsor of the publication.

Nijmegen, June 1960 J. A. B.

CONTENTS

A NOTE ON SOURCES

This work deals with the diplomatic activity brought about by the Belgian and Polish revolutions during the years 1830 and 1831.

Of the extensive literature dealing with the first problem I mention only the studies by Fl. de Lannoy, W. von Franqué, and C. Smit, the last-mentioned work being based on printed sources only. The French viewpoint is presented by E. de Guichen, and, more adequately, by R. Guyot, while the writer of a recent monograph on Palmerston's policy, Sir Charles Webster, used the invaluable private papers of the British Minister which had not so far been available.

The diplomatic history of the Polish movement of 1830 is yet to be written. There are important contributions by J. Dutkiewicz with regard to the Polish policy of Austria and France, and by M. Kukiel, who examines the outbreak of the revolution. R. F. Leslie's recent account of the Polish revolution and its social implications hardly touches on the diplomatic problems involved.

The only attempt to connect the Belgian and Polish events of 1830 together is made by Mme Perelman-Liwer; in her study Belgian internal affaires are by far the principal subject, while British sources are almost entirely disregarded.

With regard to the sources of the period, one has to bear in mind that Belgian affairs were a public concern of the Great Powers since their intervention had been requested by the King of the Netherlands; Polish affairs were not, for the Emperor-King made no request of this kind. Hence, in France and in Britain, but especially in Britain, little information is to be expected in official correspondence about the attitude of the respective governments with regard to the war on the Vistula. I have used the records of the Foreign Office and of the Ministère des Affaires Etrangères, being aware that there is little relevant information to be found in them.

Of the chief printed sources I would mention the Dutch *Gedenkstukken* which contain a selection of diplomatic reports of the period as well as semi-official and private correspondence of outstanding personalities in the Low Countries. With regard to the latter I have almost invariably found that the editor omits the portions of documents which refer to Poland – mainly accounts of military events – which are to be found in almost all documents written by the Dutch representatives in London. Pallain's reprints of Talleyrand's reports from London are necessary, as also the Russian *Recueil* by Martens; the latter unfortunately consists for the greater part of précis of documents. In view of the scarcity of Russian sources, the letters of Nicholas and the Grand Duke Constantine, his elder brother, printed by the Société Impériale d'Histoire de Russie, are quite invaluable.

For more relevant information I have referred to memoirs and the publications of private correspondence; I accumulated some evidence from English and French sources of this kind. The correspondence of Grey with Mme Lieven proved very valuable. For German material of this kind I relied on secondary works, except for a few occasional references, as this problem had received full treament in Franqué's monograph on Luxemburg; I have used, however, the publications of Metternich's and Gentz's archives.

My research for relevant archives was done almost exclusively in private collections in England and in Holland. The Lieven MSS at the British Museum proved to be amine of information, as were the Bagot MSS at Levens Hall, though the latter relate much more closely to the Netherlands than to Poland. Much was found in the Broadlands MSS of Palmerston and in the Howick MSS of Grey. The Granville MSS contain interesting letters from Lord Holland. The Wellesley MSS of the first Lord Cowley, at the Public Record Office, contain no item for the period 1830-31 referring to Vienna. The Peel, Aberdeen and Wellesley (of the first Duke of Wellington) MSS yielded no information for my purpose, nor did the correspondence between Sir Frederick Lamb, from 1832 Ambassador in Vienna, and Palmerston, kept in Panshanger MSS in Hertfordshire County Record Office. Letters from Mme Lieven to Lady Cowper, part of which are already published, in the Lamb MSS at the British Museum contain no item of importance.

In Holland I examined all the private collections of the period available at the Algemeen Rijksarchief in The Hague, as well as the Van der Hoop-Van Zuylen correspondence in the MSS of the Royal Library. The Enghuizen Collection in Arnhem yielded no information with regard to the Netherlands Minister in St. Petersburg, Baron Van Heeckeren tot Enghuizen. The most important evidence in Holland is found in the semi-official correspondence of the Dutch representatives at the London Conference with the Minister for Foreign Affairs (designated Conferentie Londen, B.Z. Nos. 1695 and 1696).

In France I tried unsuccessfully to obtain access to the Lieven-Guizot correspondence which I understand is now being edited. The MSS of the Bibliothèque Polonaise in Paris contain some material relating to international relations, including the papers of the Grand Duke Constantine, seized in Warsaw.

My research in the Lieven MSS resulted in my obtaining some material relating to the Orangist movement in Belgium; part of it is included in the present work as relevant to the problem, since I believe that Lieven's conscientiousness would have produced some results for the Orangist cause if his efforts had not been paralysed by his fear of an Anglo-French agreement directed against Russian interests in Eastern Europe.

J. A. B.

SOURCES

I. UNPUBLISHED

A.é.	Archives du Ministère des Affaires Etrangères, Paris. The following volumes were used: Angleterre 632-635; Autriche 413-415; Pologne (Consulat de Varsovie) 1-3; Prusse 274-277; Russie 181-183
B.P.	Manuscripts of the Bibliothèque Polonaise in Paris, Vols. I-IX (Cat. Nos. 350-358). Papers of the Polish Mission in Paris in 1831, also some transcripts of the papers of the Grand Duke Constantine, seized in 1830.
Bagot MSS	The Papers of Sir Charles Bagot, G.C.B., at Levens Hall. Vols. VIII-IX of the "Netherlands Correspondence" and "Letters received" (1830, 1831, 1832) were used.
F.O.	The Papers of the Foreign Office at the Public Record Office, London. General Correspondence: Austria 222-223, 227-230; Belgium 4-7; France 416-417, 420-432; Holland 175-177; Prussia 163-164, 167-175; Russia 187 and 191-193
Granv MSS	The Papers of Granville Leveson-Gower, Viscount (1833 1st Earl) Granville, at the Public Record Office, London.
Heytesbury Papers	The Papers of William A'Court, 1st Lord Heytesbury, at the British Museum, Add. MSS 41560–41562
Ellenborough MSS	The Papers of Edward Law, Earl of Ellenborough, at the Public Record Office, London.
Guizot Papers	The Papers of François Guizot, deposited in the Archives Nationales, Paris.
Howick MSS	The Papers of Charles Grey, 2nd Earl Grey at the University Library, University of Durham, Durham

Lie MSS	The Papers of the Lieven Family, at the British Museum	
	Princess Lieven to Count Nesselrode	Add MSS 47355–6
	Princess Lieven to Alexander Benckendorff	Add MSS 47415–6
	Princess Lieven, General Correspondence	Add MSS 47377
	Palmerston to Princess Lieven	Add MSS 47363
	Prince of Orange to Princess Lieven	Add MSS 47346
	Princess Lieven to her husband	Add MSS 47398
	Palmerston to Prince Lieven	Add MSS 47266
	Prince Lieven, Papers relating to the Netherlands and Italy	Add MSS 47286
	Matuszewic, Lieven, Pozzo di Borgo	Add MSS 47276–7
	Matuszewic to Princess Lieven	Add MSS 47357
Palm MSS	The Papers of Henry John Temple, 3rd Viscount Palmerston, at Broadlands.	
Panshanger MSS	The Papers of the Family Lamb, Viscounts Melbourne. The Papers of Sir Frederick Lamb (3rd Viscount) contain almost no item for 1830–1831.	
R.A.	Algemeen Rijksarchief, The Hague. Collections Falck, Verstolk and Van Zuylen were used. Where there is no further indication, reference is made to the records of the Ministry for Foreign Affairs.	

II. PRINTED DOCUMENTS AND MEMOIRS

(An abbreviation is used in the footnotes if it occurs before the full title.)

Apponyi. *Journal du Comte Rodolphe Apponyi,* 1826–1850, ed. E. Daudet, 4 vols., Paris 1913–1926.

A. Aspinall, *Three Early XIX Century Diaries,* London 1952.

Barante, *Memoirs. Souvenirs du Baron de Barante,* publiés par son petit-fils, Claude de Barante, 8 vols., Paris 1890–1901.

Belliard, *Memoirs. Mémoires de Auguste D. Comte de Belliard,* par M. Vinet, 3 vols., Paris 1842.

B.F.S.P. *British and Foreign State Papers,* edited by the Librarian of the Foreign Office, London 1833–1857 (vols. XVIII-XXXVII).

Belgium, Papers Relative to the Affairs of, presented to both Houses of Parliament by command of His Majesty, vols. A and B, London 1833.

C. Bouffin, *Mémoires et documents inédits sur la révolution belge et la campagne de dix jours,* 2 vols., Bruxelles 1912.

Broglie, A. L. V. C. Duc de: *Souvenirs 1785–1870,* 5 vols., Paris 1886.

Brougham, Lord Henry, *Life and Times, written by himself,* 3 vols, London 1871.

Bulwer. *The Life of Henry John Temple, Viscount Palmerston,* with Selections from his Diaries and Correspondence, Vols I and II edited by Henry Lytton Bulwer, 1st Lord Dalling, London 1870.

Chad. *The Conversations of the First Duke of Wellington with George William Chad,* edited by the 7th Duke of Wellington, 1956.

Correspondance. *La Correspondance de l'Empereur Nicolas Ier avec le Grand Duc Constantin,* Transactions of the Société Impériale d'Histoire de Russie, vol. 131, St. Petersburg 1911.
Ged. H. T. Colenbrander, *Gedenkstukken der Algemeene Geschiedenis van Nederland,* part X: 1830–1840. Vol. I: English Reports, vol. II: French Reports, Vol. III: Austrian, Prussian and Russian Reports, vols IV-V: Dutch Reports. The Hague 1918–1922.
Cowley. *Diary and Correspondence of Henry Wellesley, First Lord Cowley, 1790–1846,* edited by his grandson Col. H. Wellesley, London, 1930.
Dino, duchesse de, see *Revue de Paris.*
Falck. *Gedenkschriften van Anton Reinhard Falck,* R.G.P. Kleine Serie, Vol. 13. The Hague, 1913.
Brieven van A. R. Falck, ed. O. W. Hora Siccama, The Hague 1857.
Brieven van A. R. Falck, ed. O. W. Hora Siccama, The Hague 1861.
Frederick William III of Prussia. *Aus dem letzten Jahrzehnte Friedrich Wilhelm III.* Briefe des Königs an seine Tochter Charlotte, Kaiserin von Russland, ed. P. Bailleu, Hohenzollern Jahrbuch vol. 20, Berlin 1916.
F. von Gagern, *Das Leben des Gen. Friedrich von Gagern,* 3 vols., Heidelberg 1856–7.
Gedenkstukken, see H. T. Colenbrander.
Gentz, F. von, *Tagebücher,* ed. A. Fournier and A. Winkler, Vienna 1920.
Gerretson. C. F. Gerretson, *Muiterij en scheuring,* 2 vols., Leiden 1936.
Greville Diary. A Journal of the Reigns of King George IV, King William IV and Queen Victoria, by Charles C. F. Greville, Esq., Clerk of the Council to those Sovereigns, ed. H. Reeve, 8 vols., London, 1888.
Hobhouse. John Cam Hobhouse, 1st Lord Broughton, *Some Account of a Long Life,* London 1865.
Hobhouse. John Cam Houbhouse, 1st Lord Broughton, *Recollections of a Long Life,* extracts from diaries, correspondence etc, edited by his daughter, 6 vols., London, 1910.
Leopold, King of the Belgians, see Secondary Works, Th. Juste.
The Lieven Grey Correspondence. *Correspondence of the Princess Lieven and the Earl Grey,* ed. Guy Le Strange, 3 vols. London 1902.
Lubecki, Prince Franciszek Ksawery, see Smolka.
Martens, *Recueil. Recueil des traités et conventions, conclues par la Russie avec les puissances étrangères,* par F. de Martens, 15 vols., St. Petersburg 1874–1909.
Metternich. *Mémoires, documents et écrits divers, laissés par le Prince de Metternich,* 8 vols., Paris 1880–84.
Molé, Count Louis Mathieu, see de Noailles.
G. E. v. Natzmer, *Unter den Hohenzollern. Denkwürdigkeiten aus dem Leben des Gen. Oldwig von Natzmer,* 4 vols., Berlin 1887–89.
Nesselrode. *Memoirs. Lettres et papiers du Comte Charles de Nesselrode,* publiés par son fils, comte A. de Nesselrode, Paris, 11 vols, 1904–1912.
Nicholas I of Russia, Correspondence with the Grand Duke Constantine, see Correspondance.
Noailles, H. de, *Le Comte Molé, sa vie – ses mémoires,* 6 vols., Paris 1922–1930.
Nouvelle Revue Rétrospective, XV-XVI, 1901/2. Lettres du Prince de Talleyrand et de la Duchesse de Dino à Madame Adelaïde, 1830–31.
Pallain. G. Pallain, *Ambassade de Talleyrand à Londres, 1830–1834,* Paris 1891.
Revue de Paris, IV, 1923. Lettres de la Duchesse de Dino à Adolphe Thiers, 1830–1837, ed. M. Lanzac de Laborie.
A. Prokesch-Osten, *Aus den Tagebüchern,* Vienna 1909.
Aus dem Nachlasse des Grafen Prokesch-Osten. Briefwechsel mit Herrn von Gentz und Fürsten Metternich, 2 vols., Vienna 1881.

Revue des Deux Mondes, LV-LVI, 1910. Talleyrand, Sebastiani, Louis Philippe. Lettres particulières, publiées par le comte de Choiseul.
Russell. *The Early Correspondence of Lord John Russell 1805–1840,* edited by his son R. Russell. London 1913.
Sapieha, *Memoirs.* Prince Leon Sapieha, *Wspomnienia* (Souvenirs), Lwów 1912.
Smolka, St. (ed.) *Korespondencya Lubeckiego* (The Correspondence of Lubecki), 4 vols., Cracow 1909.
Stockmar. *Memoirs of the Baron Christian Frederick von Stockmar,* ed. E. v. Stockmar, 2 vols., London 1873.
Talleyrand, *Memoirs. Mémoires du Prince de Talleyrand,* avec une préface et des notes par le Duc de Broglie, 5 vols., Paris 1891–2.
Talleyrand, see *Nouvelle Revue Rétrospective* and Pallain.
Wellington, *Suppl. Despatches. Despatches, Correspondence and Memoranda of Arthur Wellesley, 1st Duke of Wellington,* in continuance of the former series, 8 vols., London 1867–1880.
The Correspondence of the late Earl Grey with H. M. King William IV and with Sir Herbert Taylor, Nov. 1830–June 1832, ed. H. Earl Grey, 2 vols., London 1867.
Zamoyski Count Andrzej, *Moje Przeprawy. Pamiętnik o czasach powstania Listopadowego* 1830–1831 (My adventures. Souvenirs of the November Insurrection) 2 vols., Cracow 1906.
Zuylen van Nyevelt H. van, *Memorandum, betrekkelijk zijne diplomatische verrigtingen in Juni, Juli en Augustus 1831* (Memorandum relating to his diplomatic mission in June, July and August 1831), privately printed 1853.

III. SECONDARY WORKS

A. v. Arneth, *Johann Freiherr von Wessenberg,* 2 vols., Vienna-Leipzig 1898.
Askenazy, S. *Polskie zabiegi dyplomatyczne w czasie powstania listopadowego* (Polish Diplomatic Endeavours during the November Rising), Biblioteka Warszawska, 1903.
Barosz, "La révolution polonaise de 1830 et la déchéance du tzar Nicolas Ier", *Annales de l'Ecole des Sciences Politiques,* 1895.
Barzykowski, St., *Historya Powstania Listopadowego* (The History of the November Rising), 5 vols., Poznań 1883–84.
Bell, H. C. F. *Lord Palmerston,* 2 vols., London 1936.
Bibl, V., *Metternich in neuer Beleuchtung.* Seiner geheimer Briefwechsel mit dem bayerischen Staatsminister Wrede, Vienna 1881.
Bosch Kemper, J. de, *Geschiedenis van Nederland na 1830,* 2 vols., Amsterdam 1873.
Colenbrander, H. T. *De afscheiding van België,* Amsterdam 1936.
Colenbrander, H. T. *Willem II, Koning der Nederlanden,* Amsterdam 1938.
Collin de Plancy, J. A. S. *Histoire de Léopold, premier roi des Belges,* Bruxelles 1835.
Corti, E., *The Reign of the House of Rothschild,* London, 1928.
Czartoryski, A. *Żywot J. U. Niemcewicza* (The Life of J. U. Niemcewicz), Poznań 1860.
Darcy, J. *Ambassade de Talleyrand à Londres,* Paris 1891.
Débidour, A. *Histoire diplomatique de l'Europe, 1815–1878,* Paris 1919.
Demoulin, R. "L'influence française sur la naissance de l'Etat belge", *Revue historique,* CCXXIII (1960).
Dinger Hattink, R. E. *De Brusselsche opstand,* Amsterdam 1930.
Dutkiewicz, J. *Austria wobec powstania listopadowego* (Austria and the November Rising), Cracow 1933.

Dutkiewicz. J. Dutkiewicz, *Polska a Francja w roku 1831* (Poland and France in the year 1831), Łódź 1950.
Franqué. *Luxemburg, die belgische Revolution und die Mächte* (Rheinisches Archiv Heft 24), von Wolfgang von Franqué, Bonn 1933.
Gleason, J. H. *The Genesis of Russophobia in Great Britain,* Princeton 1950.
Gronemann. W. Gronemann, *Die Haltung Preussens in der belgischen Frage,* Berlin 1928.
de Guichen. E. de Guichen, *La révolution de juillet 1830 et l'Europe,* Paris 1917.
Guyot, R. *La première entente cordiale,* Paris 1926.
Guyot, R. "La dernière négociation de Talleyrand", *Revue d'histoire moderne et contemporaine,* 1901–1902.
Hall, J. R. *England and the Orleans Monarchy,* London 1912.
Haussonville, M. O. d' *Histoire de la politique extérieure du Gouvernement Français 1830–1848,* 2 vols., Paris 1850.
Hillebrand, K. *Geschichte Frankreichs von der Thronbesteigung Louis Philips bis zum Falle Napoleons III.,* 2 vols., Gotha 1877–79.
Huber, G. *Kriegsgefahr über Europa,* Berlin 1936.
Jungfer, Th. *Die Beziehungen der Julimonarchie zum Königreich Polen,* Berlin 1909.
Juste, Th. *Alexandre Gendebien,* Bruxelles 1874.
Juste, Th. *Le comte Le Hon,* Bruxelles 1867.
Juste, Th. *Memoirs of Leopold I, King of the Belgians,* 2 vols., London 1868.
Kukiel, M. "La révolution de 1830 et la Pologne", *Revue internationale d'histoire politique et constitutionnelle,* Nouvelle Série, t. III (1953).
Kukiel, M. *Adam Czartoryski and European Unity,* Princeton 1955.
Lacour-Gayet, G. *Talleyrand,* 4 vols., Paris 1926.
de Lannoy. Fl. de Lannoy, *L'Histoire diplomatique de l'indépendance belge,* Bruxelles 1930.
Leslie. R. F. Leslie, *Polish Politics and the Revolution of 1830,* London 1956.
H. Lisicki, *Aleksander Hrabia Wielopolski, Margrabia Myszkowski,* 4 vols., Cracow 1878.
Martinet. A. Martinet, *Léopold Ier et l'intervention française en 1831,* Bruxelles 1905.
Masure, A. "La reconnaissance de la monarchie de juillet", *Annales de l'Ecole des Sciences Politiques,* 1892–93.
Morley, Ch. "The European Significance of the November Uprising", *Journal of Central European Affairs,* 1950.
Müller, J. *Die Polen in der öffentlichen Meinung Deutschlands 1830–1832,* Marburg, 1923.
Perelman-Liwer, F., *La Belgique et la révolution polonaise de 1830,* Bruxelles 1948.
Perniola, E. "De internuntius Mgr Francesco Capaccini en de Belgische omwenteling van 1830", *Mededelingen van het Nederlandsch Historisch Instituut te Rome,* Third Series, IV (1947).
Reid, Stuart J., *Life and Letters of the First Earl of Durham,* 2 vols., London 1906.
Richter, F. *Das europäische Problem der preussichen Staatspolitik und die revolutionäre Krisis von 1830 bis 1832,* Leipzig 1933.
Schiemann, Th. "Die Sendung des Feldmarschalls Diebitsch nach Berlin Aug.-Nov. 1830", *Zeitschrift für Osteuropäische Geschichte,* I (1910–11).
Schiemann, *Geschichte Russlands.* Th. Schiemann, *Geschichte Russlands unter Kaiser Nikolaus I.,* 4 vols., Berlin 1913.
Seton-Watson, R. W. *Britain in Europe, 1789–1914,* Cambridge 1937.
Smit. C. Smit, *De conferentie van Londen. Het vredesverdrag van 19 April 1839,* Leiden 1949.
Steinmetz, R. *Englands Antheil an der Trennung der Niederlande,* The Hague 1930.

Stern, A. *Geschichte Europas 1815–1871*, 4 vols., Berlin 1921.
Webster. C. K. Webster, *The Foreign Policy of Palmerston 1830–1841*, London 1951.
Weill, G. *L'éveil des nationalités et le mouvement libéral*, Paris 1930.
Thureau-Dangin, P. *Histoire de la monarchie de juillet*, 7 vols., Paris 1884–1892.
Tokarz, W. *Wojna Polsko-Rosyjska w roku 1831* (The Polish–Russian War in the year 1831), Warsaw 1930.

INTRODUCTION

The great strain which the Napoleonic wars brought to all governments in Europe made them agree to the necessity for the existence of a supreme institution which would watch and control the development of the international situation. The idea was embodied in the Holy Alliance of the monarchs of Austria, Prussia and Russia, and, more properly, in the Quadruple Alliance of 20th November, 1815, to which, besides these three, Great Britain was a party, and which thus constituted an association of the Four Powers which had overthrown Napoleon. France was soon admitted to the concert of the Great Powers, when the foreign troops had been removed from her territory in virtue of the Aix-la-Chapelle Protocol of 1818. Castlereagh encouraged this, as it tended to promote the return of the old European equilibrium and especially to counterbalance Russia: to this end, at Vienna, Prussia had been strengthened in the West, Norway united to Sweden, and Austria given a free hand in Italy. The three pillars of the Holy Alliance did not oppose the inclusion of France, since the legitimate dynasty had been restored in Paris.

The years following witnessed the gradual dissolution of this organization. It was Britain who kept herself away from commitments for the future, and in Castlereagh's opinion the obligations resulting from the Quadruple Alliance did not extend beyond a guarantee of the territorial settlement of 1815 for twenty years, and the exclusion of the Bonapartes from all power in Europe. France's aims were also opposed to the interests of the three.

At the Congress of Troppau in 1820 the principle of intervention in the internal affairs of the European States, should their development menace the legitimate monarchical order, was only adopted by the three Eastern Governments; it had been previously resolved at Aix-la-Chapelle that intervention was only possible with the consent of the State on the territory of which it was to take place. The French intervention in Spain in 1823, though it had the consent of Vienna, Berlin and St. Petersburg, aroused indignation in Britain. In the Portuguese crisis of 1826 Britain mastered the situation according to her own wishes.

It was the Greek question which at one and the same time united the Powers and became an apple of discord. Almost everywhere in Europe there was sympathy with the Greek struggle for freedom, which subsisted in difficult circumstances from 1821 onwards and commanded varying kinds of support from peoples and governments. Next to the principle of legitimacy, the Greek question involved that of the dissolution of the Ottoman Empire and access to the Straits leading to the Black Sea. The chief rivals here were Great Britain and Russia, the latter apparently in a stronger position to work her will owing to her direct contact with the Turkish frontiers and her connections of race and religion with the peoples of the Balkans.

When soon after his accession, in 1826, Nicholas threatened Turkey, claiming to protect the Balkan principalities, Wellington was sent to St. Petersburg to negotiate the means of affording freedom to the Greeks without, at the same time, letting Russia advance southwards too much. By the Anglo-Russian agreement of 4th April, 1826, the recognition of some sort of independence of Greece was established for the first time; while Russia admitted on the one hand British mediation between the Sultan and the Greeks, she promised not to seek advantages in Turkey, should she be required to support such mediation. Yet Nicholas succeeded in imposing upon Turkey the Treaty of Akerman in October, 1826.

On the initiative of Britain the negotiations in the Greek question were extended by the admission of France, and, after the tripartite Treaty of London, of July, 1827, which provided for an offer, or rather, an enforcement of mediation, a fleet of the three Powers was sent to the Mediterranean; and in an almost accidental engagement in the Bay of Navarino the Turkish fleet was completely destroyed in October of the same year. Action once undertaken, it was continued by the landing of the French in Morea, and by Russia starting hostilities in the Balkans in 1828. It was only the appearance of the Russian army at Adrianople – after a hard-fought campaign in which the Russians suffered great losses – which brought the Sultan to terms in September, 1829, making him agree to the independence of Greece, previously established by the Three Powers at their London Conference in March.

Throughout the negotiations Metternich vainly attempted to step in.

In the solution of the Greek question Russia played a far more important rôle than that which had been intended for her by the agreement of 1826. It was Britain which had brought France into the negotiations to counterbalance Russian influence, and yet in the advanced stages of the negotiations, in 1829 and 1830, quite cordial relations existed between Paris and St. Petersburg, to the detriment of British policy. At the time of the Turkish war France even attempted to bring forward a fantastic plan for re-modelling Europe, whereby the King of the Netherlands would be "transferred" to Constantinople, and his former possessions divided between France and Prussia. The Algiers expedition of 1830 was encouraged by Nicholas and disapproved of by Britain.

While diplomacy was engaged in the affairs of a distant corner of Europe, the liberal movement, more or less suppressed in every state of Europe for the previous fifteen years, matured sufficiently to change the face of politics and to shake the existence of the Quadruple Alliance.

The outbreak of revolution in Paris in July, 1830, initiated a series of other similar movements, of which two, one in the Southern Netherlands and the other in Poland, were destined to engage general attention during several months. The union between the Southern and the Northern Netherlands, and the establishment of the Kingdom of Poland were both included in the Treaty of Vienna, but in the first case a change in the existing state of things presented no direct threat to any possessions of the Great Five, besides their interests in the question were opposed; consent to the establishment of the Kingdom of Poland, in 1815, was only grudgingly given by the two co-partitioners of Poland, Prussia and Austria, as being the only means of preventing a simple annexation by Russia of what had been the Grand Duchy of Warsaw. Through a passing mood of Alexander it became possible for the Kingdom to obtain a more liberal constitution than that of France,[1] which tended to strengthen its separate existence and made the link with Russia still more difficult to maintain. The interest of the two

[1] In the France of Charles X there were 80,000 electors, while in the Kingdom of Poland, with a population less than one-fifth of that of France, there were some 100,000. G. Weill, *L'éveil des nationalités et le mouvement libéral*, 1930, p. 124.

Western Powers in the existence of Poland hardly went beyond sympathy, though in the establishment of an independent Polish State Castlereagh would have seen a desirable weakening of Russia and it seems that, at Vienna, his appeal for consideration for Polish nationality was made solely to that purpose.

The new régime established in France, a constitutional monarchy under Louis Philippe, was recognized by the neighbouring states within a few weeks. No plan to restore the Bourbons was contemplated, but a new wave of revolutions was feared, and it seemed likely that France would support them. The representatives of the two Powers most opposed to such a course, Metternich and the Russian Vice-Chancellor, Count Nesselrode, who happened to be spending their holidays in neighbouring places in Bohemia, drew up, at the first news of the revolution in Paris, the famous agreement, the "chiffon de Carlsbad", which stipulated that while no intervention in France would be undertaken, French intervention elsewhere would not be endured. It was then understood that the Imperial Government of Russia also reconciled itself to the new state of things in France.

The disturbances in Brussels followed closely on those in Paris. It was believed at first, however, that no serious consequences were to be anticipated. The King of the Netherlands received a delegation from this southern capital and summoned the States General which on 13th September pronounced the advisability of the administrative separation of the two parts of the Kingdom. Before giving effect to this decision the King wished to restore order. His attempt failed in 4 days' fighting in Brussels, 23rd to 27th September, and the King applied to the Powers of the Quadruple Alliance for assistance.

There seemed to be little likelihood of the Powers embarking upon a military expedition on his behalf. The most friendly monarch, Frederick William III, trembled for his Rhenish provinces, and Austria had to reckon with the influence of the French revolution upon the oppressed Italians. Though neither Prussia nor Austria was a constitutional state, their governments – both the sovereigns and the principal advisers of the Crown – were mindful of the consequences of harsh measures. The influence of public opinion, however small, and barely allowed to express itself through the Press, was taken into account. The great agitation which prevailed in some parts of the Prussian and Austrian dominions after the Paris events, together with serious disturbances in some of the minor German states, made the two governments anxious to avoid the application of force in that part of the former French possessions which, after July, France seemed least likely to have forgotten. Lastly, Britain, which had showed a disinclination for intervention before, was – apart from all other considerations of both foreign and internal policy – unable to intervene on her own.

There was no time for delay. Before any news could be obtained from the Russian capital, the other Powers offered mediation, and a Conference, from which Great Britain thought it impossible to exclude France though William I's appeal was not addressed to her, was called to assemble in London on 4th November. This was a meeting of the representatives of two constitutional states and three absolute monarchies, who had to agree in working out a solution of the conflict between a constitutional sovereign and his revolted subjects in an area where moreover many international interests had been involved for centuries.

Of all the parties Russia was the least interested in the affairs of the Low Countries and yet most opposed to the solution proposed. Her ruler, Nicholas I, was a convinced believer in the divine right of princes to govern as they pleased, and he fully applied this principle in his own empire. He possessed little personal culture: all his interests were concentrated on military affairs. Though he had, before his accession, travelled in Western Europe, he hardly understood the enormous difference between his own subjects and other peoples; aware of the existence of constitutional forms of government he could not see a constitution in any other light than as a voluntary cession of some of his rights by a sovereign. Born in 1796, Nicholas never knew until 1823 that he would occupy the throne, and even after that time he was not introduced to the affairs of state. His accession was preceded by a military conspiracy which he mercilessly suppressed.

He had consequently shown great reluctance, which was only increased by the news from Brussels, in recognizing Louis Philippe. The Netherlands Crown Prince, the Prince of Orange, was his brother-in-law and a personal friend. And though Nicholas had recognized Louis Philippe, this act was soon followed by the widely publicized mobilization of the Russian army.

It seemed, therefore, that the London Conference would not

be kept together for long, the more so as, on the one hand, the development of affairs in the Southern Netherlands did not tend to make its task any easier, and on the other, the attempt at a solution of such difficulties by means of an ambassadorial conference was a new venture, and therefore unlikely to succeed in what seemed to be a very complicated problem. For the Powers of the Quadruple Alliance the creation of the Kingdom of the Netherlands had provided a barrier against France, and to this end a line of fortresses had been erected along its Southern frontiers. It was clear that were the ties between Belgium and Holland loosened or severed, the fortresses would pass, if not to France herself, at any rate into the care of people most likely to be her friends. But the demolition of the fortresses was, to the French Government, the *sine qua non* achievement to follow the dissolution of the Kingdom of the Netherlands; it was of such paramount importance that Louis Philippe announced their demolition in his speech at the opening of Parliament in July, 1831, though no public act of the Powers made such an announcement possible: the April protocol of the Four Powers in London on the subject had been communicated confidentially to the French representative, but it was not before December of that year that a convention of the Four Powers with the King of the Belgians provided for the demolition of some of the fortresses without – as was only natural, but much to the damage of the prestige of the Paris Government – any French participation in the proceedings.

Lastly, one more factor indicated difficulty in continuing the Conference: its unhomogeneous composition. Here the balance favoured France. The French representative, Talleyrand, whose appointment to the London Embassy was one of the first public acts of the French King, possessed, apart from more than ordinary abilities, an experience second to none in the Conference. Shortly after the opening of the Conference the change of government in Britain brought in another important personage, Palmerston, of whose qualities Talleyrand also was soon aware. Both these men would, for different reasons, be at times opposed to the interests of the Belgians; but they would always object to unequivocally favouring the King of the Netherlands. Of those from whom the most stubborn opposition was to be expected one, Prince Lieven, the Ambassador of Russia, was a nonentity – and all the members of the conference were aware of this: in the eighteen years of Lieven's

London Embassy there was ample opportunity to arrive at this conclusion even for those who only met the Russian Ambassador outside England; the other, Baron Bülow, Minister of Prussia, was tied down to his instructions, which required him to avoid a conflict at all costs. Personal factors influenced the activities of the remaining members of the Conference to a greater degree than could be expected, to the detriment of the policies followed by their Courts. Wessenberg, the second Austrian representative, "did not like the Belgians",[2] but as an able diplomat he contributed much to bringing together opposing standpoints in the Conference; Matuszewic, Lieven's deputy, when in opposition, which on account of his abilities could be considerable, was unwillingly half-hearted, being convinced that the wishes of The Hague, whether corroborated by his instructions or not, were not realizable.[3] The Austrian Ambassador, Prince Esterhazy, remained grandly detached. It remains to be noted that a not inconsiderable influence was exercised in London by the spirited wife of the Russian Ambassador; it did not, however, penetrate the intricacies of the Conference's business, and whatever her influence with Grey – and it does not appear that Grey allowed himself consciously to be so led in any public question – it seems to have been next to none with Palmerston.

After their successful defence of Brussels against the Royal troops the Belgians elected their national assembly, the Congress, proclaimed their independence, and, contrary to advice sent even from Paris, voted late in November the deprivation of the House of Orange from all its powers. This was a tremendous challenge to the three Eastern Powers who represented the principle of legitimacy. However, before they could resolve on a reaction, to which Nicholas wanted to give a decidedly military character, the outbreak of the revolution in Warsaw on 29th November drew their attention elsewhere. The initiative now passing unequivocally to the Western representatives, that is to Talleyrand and to the new Whig Foreign Secretary, Palmerston, the Conference declared on 20th December that in future Belgium should exist as an independent State, and on 20th and 27th January, 1831, it set out in its Protocols the bases

[2] F. de Lannoy, *L'histoire diplomatique de l'indépendance belge*, 1930, p. 55.

[3] Cf. especially Matuszewic to Nesselrode, 21 Oct. 1831, H. T. Colenbrander, *Gedenkstukken der Algemeene Geschiedenis van Nederland 1795–1840*, Part X, Vol. III, 496.

for the separation of Belgium and Holland. Additionally encouraged by this result in the international field, the Belgians proceeded to elect the Duc de Nemours, younger son of Louis Philippe, as their king; but as this was unacceptable to most of the Conference Powers, they contented themselves with a Regent.

The King of the Netherlands accepted the January Protocols, but the Belgians did not, and, contrary to the Conference's stipulations, throughout the earlier half of 1831 they claimed the whole of Luxemburg, Limburg and Zealand Flanders (the left bank of the Scheldt); the Regent publicly challenged the Conference Powers, the German Federation and the King of the Netherlands by declaring that Luxemburg was a part of Belgium and would remain so.

It should be noted, on the other hand, that the acceptance of the "Bases de séparation" by The Hague had given rise to false interpretations outside Holland. Even St. Petersburg was *au courant* of the general desire in the northern part of the Kingdom of the Netherlands for a separation from Belgium. But this never meant that Holland – and least of all the King – was prepared to treat the Belgians as equal partners. Moreover the Conference Powers were themselves not clear about the rôle of the Conference: should it mediate between the two parties concerned or enforce the decisions taken in London upon them – the latter course certainly inadmissible to the Eastern Powers – or put pressure upon the Belgians only – difficult and dangerous when Paris was to be taken into consideration. The Powers at least tried to achieve an appearance of impartiality by refusing to admit the Netherlands plenipotentiaries to the Conference room, in flagrant violation of the Aix-la-Chapelle protocol and of common sense, which would require the country demanding intervention in its affairs to be represented. This in itself would be sufficient argument for William I to oppose the Conference; if his political sense equalled his other undoubtedly exceptional qualities, he might be able, by withdrawing from the Conference, to save his dignity at least with regard to his most faithful ally, the Emperor of Russia, who moreover, at this time, to most observers seemed about to free himself from the complications of the revolution in Poland. However, the ambiguous, half-hearted and at times petty way in which William's two plenipotentiaries, the London Ambassador Falck and the second delegate, Baron van Zuylen van Nyevelt, Ambassador to Turkey, were trying to influence the decisions of the Conference in his favour – of which the protocol of 27th January on the subject of the debt is outstanding evidence – easily created the impression with the other Powers that the King had acquiesced in the *fait accompli* of the loss of his sovereignty in Belgium. Palmerston, at any rate, soon reached the conviction that Falck followed the King's instructions very unwillingly. All the more incomprehensible, to the Foreign Secretary as well as to the British Ambassador at The Hague, Sir Charles Bagot, was William I's insistence on his sovereign rights in his Southern capital, as strong in the beginning as at the end of 1831, and at either time seemingly out of touch with the reality in the Low Countries.

Not until May, however, did the Conference propose suspending relations with the provisional Government of Belgium as a means of bringing it to terms; even so their representatives at Brussels delayed the measure. It had been thought that a king might stabilize the situation in Belgium, and after a vain search over some months the person of Prince Leopold of Saxe Cobourg, in many ways agreeable to those concerned, offered such a guarantee; but it appeared that nothing would induce the Belgians to adhere to the January Protocols. Encouragement was given to them not only by French, and, to a lesser degree, by British public opinion, but also by the French Government which throughout February and March refused its consent to agreements signed by Talleyrand; and later their resistance was also stimulated by the prolongation of the revolution in Poland.

The Polish revolution was, unlike the Belgian, essentially a military affair. "Au lieu des émeutes bourgeoises de Bruxelles ou des conspirations à huis clos du carbonarisme italien, c'étaient de formidables batailles..." involving "...des armées de cent mille hommes." [4] This was a war which for over half a year engaged the attention of foreign military observers.

Although there were many reasons for dissatisfaction in the "Congress"-Kingdom of Poland, the leading personalities never seriously considered recourse to arms as a means of obtaining an improvement. The movement began with a military conspiracy, and its outbreak was almost accidental; moreover it greatly embarrassed both the senior officers and the officials of the Government. When order was re-established, the Grand Duke Constantine,

[4] P. Thureau-Dangin, *Histoire de la monarchie de juillet*, 1884–1892, I, 166.

Commander-in-chief of the Polish army, who during the attack on his palace narrowly escaped with his life, was allowed to depart with the Russian regiments; in the negotiations which preceded his departure he uttered a promise to intervene with the Emperor and King in order to obtain mercy for the guilty as well as to bring about an improvement in the governmental practices. This aroused curiosity as Constantine's behaviour was generally considered, abroad especially, as the main cause of the military rising. The Polish Diet was summoned for the middle of December and in the meantime dictatorial powers were entrusted to General Chłopicki, then in his late sixties, a popular and able soldier, but a convinced opponent of war with Russia. Soon after its first meeting the Diet declared that the events in Warsaw initiated a national movement justified by the violation of the constitution,[5] but no offensive action was undertaken by the troops and no attempt made to rouse Lithuania, although the return of the Polish provinces which had been incorporated in Russia, was claimed by public opinion and constitued one of the aims of the revolution. It was not before the return of the delegation sent – even before the meeting of the Diet – to the residence of Nicholas that hopes of a peaceful settlement had to be entirely abandoned: Nicholas would not hear of anything but a submission, as demanded in his proclamation of 17th December, which included only a very general promise of pardon and no reference to any political measures in the Kingdom. The Diet replied a few days later by voting, on 25th January, the dethronement of the House of Romanov, a declaration regretted by friends and enemies of Poland abroad, as well as by many Poles themselves.

The Russian army under Field-Marshal Diebitsch, the victorious commander in the late Turkish war, entered the Kingdom early in February, and, after two minor Polish victories, the extremely bloody battle of Grochów, fought on 25th February in the vicinity of Praga, the right bank suburb of Warsaw, seemed to have brought the revolt to an end. The Poles fell back upon Praga, and during the night crossed the river into Warsaw. Owing to the indecision of the Russian commander and the great losses of his army, the Poles were not pursued. After an unsuccessful attempt to persuade them to submit, Diebitsch retreated eastwards.

[5] This declaration, under the title of *Manifesto of the Polish Estates,* is translated into English in the *Annual Register* for 1831, Chronicle, pp. 407-414.

During the winter months of 1830/31 the end of the Polish movement was generally anticipated, either brought about through negotiations, as seemed possible at first, or through a military defeat – hence the engagement at Grochów was taken for granted as a decisive Russian victory from the first news of the battle. Yet under the influence of the Polish spring offensive and of reports from Lithuania, from April onwards the events on the Vistula presented a different picture to the Western Governments. At the first news from Warsaw in December, 1830, the French Government was already in an embarrassing position with regard to Poland: Louis Philippe desired peace in order to stabilize his external and internal situation, but public opinion was loud in condemning the inactivity of the Government, the more so as it was generally believed that the Polish revolution was the main obstacle which prevented Nicholas from attempting to restore the old order in France. Nor were the French satisfied with the Government's Belgian policy: if a "reunion" of Belgium to France, as the extremists claimed, was not practicable, France should unconditionally support the Brussels authorities against dictation by the London Conference. But apart from confidential soundings in London nothing was done by the Government with regard to Poland, and, if the Cabinet of Laffitte, in power until March, showed a certain leaning towards the Poles, the improvement of their military position coincided with the appointment of the new moderate premier in France, Casimir Périer.

English society sympathized with the Poles, but the Government professed itself unable to favour them politically as this would inevitably lead to a breach with Russia and perhaps with the other two Eastern Powers; being opposed to France over the Belgian question, Britain could not afford to lose their support. Nevertheless, some representations concerning Poland were made at St. Petersburg by both Western Governments.

Austria and Prussia were opposed to the Polish revolt, both on account of their Polish provinces and of the repercussions in Germany, of which there was ample evidence. But they were not likely to act militarily: Vienna had to watch developments in Italy, which, especially during February and March, 1831, produced a great danger of war with France; and besides, since the Greek crisis, Metternich was not easy about his relations with St. Petersburg. The Prussian Government still feared for the fate of its

Western possessions, being moreover vitally interested in the ultimate settlement of the frontiers of Belgium, where important strategic positions might fall in the hands of its potential enemies.

The agitation in Belgium and the fear of annexationist ambitions of France brought the Conference in June to concessions for the benefit of the Belgians: the "Bases de séparation" having been abandoned, a new project was drawn up which provided for the possibility of assigning Luxemburg to Belgium. This project – the XVIII Articles of 25th June, 1831, – was accepted by the Belgians, and Prince Leopold, having shortly before accepted the Belgian crown, went to Brussels in July.

After the completion of these negotiations the main obstacle to a more active interest of France and Britain in the affairs of Poland seemed to have been removed. The Poles had in the meantime sustained a major defeat at Ostrolenka on 25th May, but the Russian army was suffering considerably from cholera, lack of provisions and desertions; early in June Diebitsch died. The continuance of Polish resistance was in itself an argument for Polish success and it was supplemented by the rising in Lithuania, the extent of which remained fairly obscure. Following constant representations by Polish agents, who, though not enjoying diplomatic status, were in contact with Western Ministers, both Britain and France protested in Berlin against Prussia's breach of neutrality in the Polish war by help of various kinds afforded to the Russian army. Similar protests were made in Vienna with regard to the internment of a Polish corps which took refuge on Austrian territory. These representations, however, carried little weight since they were accompanied, or followed, by the news that the French *démarche* to Britain, concerning a joint diplomatic intervention at St. Petersburg in order to bring about a suspension of hostilities on the Vistula, had been refused. Until that time, Berlin, St. Petersburg and Vienna had watched with obvious anxiety the concern of the Western nations over the Polish war, and feared an intervention, perhaps of a more serious character.

Before it could be judged whether this refusal was absolute, the situation in the West changed entirely: the King of the Netherlands, who long before had shown offence at the lenient attitude of the Conference towards Belgian resistance, appeared to be freshly stimulated in his annoyance by the concessions made to the Belgians as well as by the usurpation of the throne in what he considered to be the southern part of his dominions; on 2nd August the Royal troops renewed hostilities. Had it not been for the French armed intervention, with great reluctance approved by the Conference, Belgium might well have been overrun by Dutch troops within a few weeks. Since, rightly or wrongly, the French considered themselves bound to protect Belgium by the prolonged presence of their forces in that country, new problems were created both for the French and for the British Government. At the same time, the Polish movement was slowly expiring. The military action was brought almost to a standstill – partly at least under the influence of news from France of an alleged diplomatic intervention, which would make an unsuccessful military engagement especially dangerous – and the whole movement was virtually brought to an end by the taking of Warsaw on 8th September, 1831.

Almost simultaneously the French gave way in the matter of withdrawing their army from Belgium. The project of a settlement between Holland and Belgium was then revised once more; it turned in favour of Holland when compared to the XVIII Articles of June, though it was much less favourable to this country than were the January "Bases de séparation". This new project, the XXIV Articles, was accepted by the Belgians conscious of their defeat, and the Conference signed the Treaty with the King of the Belgians on 15th November, 1831. But the King of the Netherlands was still not disposed to agree, and it soon appeared that for this reason the Emperor of Russia would not agree either. On hearing of St. Petersburg's decision both the Austrian and Prussian Governments withdrew their consent; the situation resembled that of a year earlier, when the two Powers, but especially Metternich, regretted their early recognition of Louis Philippe after learning that Nicholas was against it. Eventually, by the time prescribed, only France and Britain presented their ratifications.

The object of the present work is to describe the consecutive relevant stages of the international situation, as far as the two revolutionary movements, in Belgium and Poland, were concerned, and as far as the one influenced the attitude of the Powers towards the other. No comprehensive survey of events in either country is given, not even of the diplomatic history of the two movements: these are subjects in themselves and the reader is referred to the note on sources and to the bibliography. As far as Poland is concerned, no attempt is made to correct all the diplomatic information quoted, unless it is too obviously in conflict with the facts as known: it was this information, correct or not, which inspired the movements of the diplomats. Similarly, no reference is made to the Press, which again would involve research out of all proportion with the scope of this study. Since, however, there is no comprehensive monograph on the diplomacy and on the repercussions abroad of the Polish revolution, some general observations are necessary.

The Press in France and in Britain, as well as in Germany and elsewhere, was pro-Polish, and therefore likely to exaggerate any information in favour of the Poles. It seems that it was quite a powerful factor in their favour, since there was little [6] or no official and reliable information *from* Poland. On the other hand, there was plenty of information *about* Poland. During 1831 news concerning Poland constituted the greater part of diplomatic reports from Vienna, Berlin and St. Petersburg; this was especially true in the case of Chad, the British Minister in the Prussian capital. It is impossible to assume that the private opinions of the writers did not influence the selection of the sources on which they based their reports.

Since all diplomatic activity with regard to Poland had to be very confidential, as the Kingdom was Nicholas' concern only, the personal relations of those who might strongly influence the proceedings one way or the other were of a more than usual importance. A review of the people concerned may conveniently start with Berlin. Of all the Powers Prussia was most interested in a solution of *both* the Polish and the Belgian questions; one reason was her common frontiers with the two countries. The King, whose chief aim, after more than thirty years of a troublesome reign, was to keep clear of all troubles, might well be expected to lend a willing ear to propositions tending to promote peace. It seemed even possible that he would be prepared to sacrifice his new Rhenish provinces to this end. "Si les Français ne vont que jusqu'au Rhin", he is reported to have said, "je ne bouge pas." [7] If this was not exactly the view of his closest advisers, there was a war party in Prussia as well, and Ancillon, the Under-Secretary for Foreign Affairs, seemed to side with it. Members of the diplomatic corps complained about the

[6] There were Prussian, French and Austrian consuls in Warsaw.
[7] A. Débidour, *Histoire diplomatique de l'Europe, 1815–1878,* 1919, I, 275.

effects of the illness of the more statesmanlike Minister Bernstoff, for whom Ancillon frequently deputised, withholding at times quite simple and otherwise available information. The French and British missions in the Prussian capital were led by men of little influence, Mortier, the nephew of the Napoleonic Marshal, and Mr. George William Chad, a country gentleman brought into the diplomatic service by the Duke of Wellington, who himself had no very high opinion of his protégé.[8]

When Palmerston took over the Foreign Office, he and Grey immediately began to plan the removal of the Tory diplomats, but the necessary delay in effecting this change lasted beyond the year 1831. Lord Heytesbury was kept in St. Petersburg at Nicholas' request, and it will be seen that Palmerston believed that this might be more effective in Polish affairs than to bring in a man perhaps not entirely to the taste of the Russian capital. Yet both Heytesbury and Lord Cowley, the Ambassador in Vienna, the former perhaps not versatile enough for the delicate task assigned to him with regard to Poland, showed none of the aversion which the most liberal Whig Minister Lord Holland kept for the Belgians; on the contrary, Heytesbury's aversion to Russia would have made him sympathetic toward the Poles,[9] had it not been for his concern with the Belgian policy of France.

French diplomats in the three Eastern capitals were not likely to achieve much for the Polish cause. To begin with, being a Frenchman from Orleanist Paris in itself constituted a liability. Maison, who had to leave the Foreign Ministry having hardly taken office, only succeeded, through his colleague at Constantinople, in creating a major diplomatic incident. Mortemart on a special mission to Nicholas was anxious to avoid war, but he and Flahaut, on a similar mission in Berlin, might have done better, were it not for the renewal of hostilities in the Low Countries. Bourgoing and Mortier were *Chargés d'affaires,* without special position; the former, moreover, was much impressed with Russian strength.

All these considerations are to be borne in mind in an examination of what appears to be the diplomatic failure of the Polish

[8] Little is known about Chad. Cf. *The Conversations of the First Duke of Wellington with George William Chad,* edited by the 7th Duke of Wellington, 1956.

[9] Cf. Heytesbury's opinion on the Polish revolution in his dispatch of 20th December, 1830, F. O. Russia; also Mortemart to Sebastiani, 4 June, A.é. Russie.

revolution. A number of facts already established may be summarized beforehand. There existed a strong tendency at St. Petersburg to press at almost all costs the legitimate cause of the House of Orange in Belgium. The outbreak of the revolution in Poland led to the first step towards Belgian independence: the London Protocol of 20th December, 1830. Though tied up for the time being by Warsaw events, the three Eastern Powers maintained a fairly strong line as regards Belgian developments throughout the winter of 1830/31, supported by the general anticipation of a Polish defeat, but above all, by Britain; Britain had been opposed to armed intervention, but now she gave the lead in opposing the French candidate to the Belgian throne. But the prolongation of the revolution and some successes of the Poles increased the intractability of the Belgians, and made the Eastern Powers unable to prevent concessions being made to the Brussels authorities. Contrary to expectation, the King of the Netherlands had recourse to arms in defence of his rights; the Conference summoned at his request had abandoned his cause. William's determined action prejudiced both the willingness and the ability of the more or less friendly Western Governments to try to prevent the defeat of the Poles, which followed in September, 1831. After the termination of the Polish contest the Eastern Powers adopted a stronger attitude and contributed much to the advantage of the Dutch in the new negotiations, but the King of the Netherlands refused his consent to the new proposals. It was known that Russia was very much exhausted by the Polish war and would be unable to support William I's claims. But Nicholas not only considerably delayed his ratification of the Treaty, but also made Prussia and Austria delay theirs. Circumstances created by the negotiations on the fortress question were such that this delay alone – seen, moreover, as a refusal – caused much irritation in London and Paris.

CHAPTER I

RUSSIA, THE RECOGNITION OF LOUIS PHILIPPE AND THE BEGINNING OF THE BELGIAN REVOLUTION

After the establishment of the Orleans monarchy in Paris Britain was the first of the great Powers to recognize the new state of things in France; her intention to do so was known to the members of the diplomatic corps in London about August 20th.

The three Powers of the Holy Alliance had to choose between following the example of London or maintaining the principle of legitimacy. As to Prussia, Frederick William III was most interested in the tranquillity of his Western provinces, and to him the government of Louis Philippe represented a guarantee of stability in France. The Austrian Chancellor, being himself opposed to the recognition of the revolutionary throne, seemed to have yielded to Nesselrode's opinion, but he regretted this afterwards.[1] A somewhat similar situation could be observed in London about this time: although Bülow, the Minister of Prussia, and Matuszewic, the *Chargé d'affaires* of Russia, had expressed their belief that their governments would recognize the new régime in France, they withdrew their statements on hearing of the hesitation of Prince Esterhazy, the Austrian Ambassador.[2]

Within the first decade of September Louis Philippe was recognized by his most important neighbours; he had yet to obtain the *placet* of the Emperor of Russia. The uneasy proceeding which led to this step, as well as the fairly hostile attitude of Nicholas towards the King of the French, in consequence of the Belgian events, is the subject of the present chapter.

I

The heads of the Russian missions in Paris and London were very much inclined to promote the establishment of friendly relations

[1] Metternich to Nesselrode, 1 Sept., Nesselrode to Metternich, 5 Sept., *Lettres et papiers du Comte Charles de Nesselrode*, 1904–1912, VII, 151 sq.

[2] Falck to Verstolk, 27 Aug., Falck Papers, R. A.

between their Court and the Palais Royal. The part played by Pozzo di Borgo, the Ambassador in Paris, has been amply described by the French historian de Guichen.[3] Matuszewic expected England to recognize the new régime in France; unlike Pozzo's arguments, his dispatch of 5th August, in which he advocated the policy of co-operation with England and warned against harsh measures with regard to the revolution, was fully approved by Nicholas.[4] Nevertheless, before that, as soon as the first news from Paris reached him, he gave orders that Bourgoing, the French *Chargé d'affaires,* should be handed his passport and he issued restrictive regulations for French subjects.[5] Pozzo was advised to leave the Embassy and to order Russians staying in France to leave the country.[6] Both the French *Chargé d'affaires* and the Ambassador, anxious to reduce tension, ignored these instructions. At the same time Lord Heytesbury, the British Ambassador, had an interview with the Emperor, in which Nicholas showed himself disinclined to enter into the French internal complications, being also aware of the general unrest in Europe. He was uneasy "for Spain, for Italy, and, above all, for Belgium." "He had received", wrote the British Ambassador, "a very uncomfortable letter from his brother-in-law, the Prince of Orange, who told him that he been taken completely by surprise and could not assemble his means of resistance in less than two months."[7]

Nicholas stressed these apprehensions in a letter to his brother, the Grand Duke Constantine; he feared that France would attempt to recover her old frontiers, and that this would create a very precarious situation. To be ready against all eventualities, therefore, some military measures had been taken.[8]

In London it was believed at this time that Nicholas would undertake nothing before consulting Nesselrode who was still in Ger-

[3] E. de Guichen, *La révolution de juillet 1830 et l'Europe,* 1917, pp. 107-138.

[4] Matuszewic to Lieven, 5 Aug., F. F. Martens, *Recueil des traités et conventions conclus par la Russie,* 1874–1909, XI, 434-6, also XV, 101 (concerning Pozzo); the latter point is stressed by Th. Schiemann, *Geschichte Russlands unter Kaiser Nikolaus I.,* 1913, III, 13.

[5] Th. Schiemann, "Die Sendung des Feldmarschalls Diebitsch nach Berlin, Sept.-Nov. 1830", *Zeitschrift für Osteuropäische Geschichte,* I, (1910–1911), 5-6.

[6] de Guichen, p. 136.

[7] Heytesbury to Aberdeen, 20 Aug., F. O. Russia, quoted by de Guichen in a French translation, pp. 155-6.

[8] Nicholas to Constantine, 18 Aug., *La Correspondance de l'Empereur Nicolas Ier avec le Grand Duc Constantin,* Transactions of the Société Impériale d'Histoire de Russie, Vol. 132 (1911), p. 36. Quoted as Correspondance.

many and who had given evidence of his pacific inclinations; having learnt of Heytesbury's report, however, the Netherlands Ambassador considered the Russian Emperor much more affected by events in Paris than it would have appeared from the attitude of his representatives, and he therefore thought it likely that the special envoy of Louis Philippe would have to wait at the Russian frontier for his permit to enter the country.[9]

A few days after Heytesbury's audience Nicholas saw the French *Chargé d'affaires*. In a long conversation Bourgoing was told that the recognition of Louis Philippe would be a matter for joint consideration with other Powers. Following Bourgoing's plea for the necessity of a more liberal régime in France the Emperor made the following statement: "Je ferai ce que je pourrai, mais n'attendez pas que je sacrifie mon honneur. Le tems, l'avis des autres cours, la tournure que les choses prendront en France, voilà ce qui pourra me déterminer. Mais je le répète, je ne transigerai jamais avec mon honneur." Shortly afterwards measures against French subjects were withdrawn.[10]

In the meantime, however, measures of a different kind were under consideration. To face a possible French expansion Nicholas thought it expedient to act in concert with his allies. For this purpose Count Orlov, his principal aide-de-camp, and Field-Marshal Diebitsch, commander of the Russian forces in the late Turkish war, were sent respectively to Vienna and Berlin.[11]

The King of Prussia apparently expected some more pronounced steps on the part of his son-in-law. As soon as the Paris news reached him he ordered a staff officer, Major von Thümen, to proceed to St. Petersburg with instructions to counteract any such steps. Von Thümen, however, failed to obtain from Prince Lieven (at that time deputizing for Nesselrode) even an affirmation that Louis Philippe would be recognized by Russia, and on 24th August the Prussian Minister in St. Petersburg reported that the mission of Diebitsch had already been decided upon. In vain did Frederick William order him to try and stop this expedition; two days later Diebitsch's instructions were signed.[12]

[9] Falck to Verstolk, 7 Aug., 7 Sept., Falck Papers, R.A.

[10] Bourgoing to Molé, 24 Aug., A.é. Russie; Schiemann, Die Sendung Diebitschs, p. 7.

[11] Nicholas to Constantine, 29 Aug., Correspondance, p. 46.

[12] Schiemann, Die Sendung Diebitschs, p. 9.

The principal document concerning this mission, the memorandum, as well as covering letters and dispatches of the same time, all underline the same threefold theme: 1. the intention not to intervene in French internal affairs 2. the intention to prevent a French intervention elsewhere 3. the desire for unity of action of the three Powers and, if possible, of England as well, and for measured language with regard to France. Should Louis Philippe ask for recognition, all the Powers would inform him that this would be a subject for joint consideration. If a negative answer were to be decided upon, this should be "une déclaration collective mais conçue dans les termes les plus concilians." Lastly, "une position particulière" of England was mentioned, coupled with the expression of an anxious hope that she would join the others in this European question.[13]

The other important document, the dispatch to Count Alopeus, the Russian Minister in Berlin, stated that the Imperial Cabinet approved entirely of the Carlsbad agreement and that they wanted to keep the unity of the three Powers in the same way as did the other two. The immediate result of this attitude was to make the recognition of Louis Philippe subject to joint deliberation; should the Powers see themselves compelled to abandon the principle of legitimacy, they would require from the King of the French a guarantee that he would recognize the territorial settlement of Vienna and the Protocol of Aix-la-Chapelle, of 1818, in return for their consent to the new state of things in France. In the meantime, he would be known as the lieutenant-general of the Kingdom. In view of the urgency and importance of the situation Field-Marshal Diebitsch was being sent to Berlin. The Emperor realized that the choice of this person might be considered as intentionally offensive, but the circumstances were precarious enough: only the conviction that a meeting of the Sovereigns themselves might produce an even worse impression made him abandon the idea of making such a proposal.[14]

[13] There is an account of the contents of this memorandum in de Guichen, p. 121, reference being made to Nesselrode to Pozzo, 15 Aug. This account coincides with that of the Carlsbad agreement. Nesselrode, who was still in Germany, may have sent his draft to Pozzo and Lieven at the same time, but it would not be final before Nicholas' approval.

[14] "Copie d'une dépêche au Comte Pozzo di Borgo en date de St. Pétersbourg le 18 Août", same to Matuszewic, and the Mémoire, in A.é. Russie; Lieven to Alopeus, 13/25 Aug., B.P.; It appears from the text that the first two are dated

The memorandum bears the date of 20th August, but Diebitsch did not leave for Berlin until 31st August, the day on which General Athalin, the special envoy of the King of the French, arrived in the Russian capital. This mission was of particular concern to Louis Philippe, who himself made alterations in Molé's draft of the letter to Nicholas in order to soften some expressions, and also chose his favourite aide-de-camp to be charged with it.[15]

The Emperor of Russia decided to receive Athalin in the capacity of a representative of the Lieutenant-General of France, the title conferred upon the Duke of Orleans by Charles X. [16] The first report of Athalin from St. Petersburg was not very promising. He was glad that his departure from France had been delayed, as the order allowing French travellers into Russia had reached the frontier post through which he had to pass only a few hours before his arrival there. A few days later he was seen by Lieven, who appeared favourably disposed towards the new régime in France, although he told Athalin that the Emperor's intention was to undertake nothing except by previous agreement with his allies.[17]

On 6th September Nicholas received the French envoy in private audience. Athalin noted "une bonté et une grâce toute particulière" shown to him during the interview, which lasted two hours. Yet Nicholas regretted that Louis Philippe did not content himself with the title of Lieutenant-General of the Kingdom. He most categorically denied any intention on his part to intervene in French affairs. He seemed pleased to hear that relations between Charles X and the ex-Duchess of Orleans remained the same as before the revolution. At the end of the audience Nicholas somewhat hesitatingly informed the French envoy that an answer would probably not be given before at least ten days.[18]

On 15th September Athalin learnt from Lieven that nothing had been decided as regarded the recognition of Louis Philippe by Russia, but that the news of the revolution in Brussels, which reached Berlin after the King of Prussia had resolved to recognize the

---

according to the old style, unlike the Mémoire. These are some of the papers of the Grand Duke Constantine, seized by the Poles after the revolution in Warsaw. See below, p. 90.

[15] H. de Noailles, *Le comte Molé,* 1922–1930, V, 91 sq.

[16] Nicholas to Constantine, 29 Aug., Correspondance, p. 46.

[17] Athalin to Molé, 1, 4 Sept., A.é. Russie.

[18] Athalin to Molé, 7 Sept., A.é. Russie.

King of the French, "est venu remettre tout en question." In Berlin, however, the hesitation appeared to be only slight. The news of Diebitsch's mission speeded the execution of the measure already decided upon.[19]

Nesselrode arrived in St. Petersburg on 22nd September. Three days later he told the French *Chargé d'affaires* that he had passed through Berlin two days after the envoy of the King of the French, Count Lobau, left the Prussian capital "emportant la reconnaissance et après avoir reçu du Roi de Prusse l'accueil le plus flatteur." Yet Nesselrode was most non-committal with regard to the possibility of a similar decision on the part of his Imperial master, to whom the decision of his father-in-law was very displeasing. The Vice-Chancellor had difficulty in explaining to Nicholas the necessity for acquiescing in the revolutionary change. Indeed, on Nesselrode's memorandum he wrote: "Je me rends à votre raisonnement, mais j'atteste le ciel que c'est et sera toujours contre ma conscience et c'est un des plus pénibles efforts que j'aie jamais fait." According to the Russian historian Martens Nesselrode would not dare risking another step so much opposed to his master's view.[21]

The first consequence of this important decision was the letter of the Vice-Chancellor to the Grand Duke Constantine in which he informed the Cesarevitch that the advice given to the French Consul in Warsaw, Durand, to take the oath of allegiance to the

[19] Mortier to Molé, 4 Sept., de Guichen, p. 173; Athalin to Molé, 18 Sept., A.é. Russie.

[20] Bourgoing to Molé, 26 Sept., A.é. Russie; A. Masure, "La reconnaissance de la monarchie de juillet", *Annales de l'école des sciences politiques,* 1893, pp. 83-5, quotes the opinion of Lobau who met Diebitsch and Nesselrode on 8 September at a dinner at Charlottenburg, after he had obtained the King's answer to Louis Philippe. Lobau ascribed to the influence of those Russian officials that Werther, the Prussian *Chargé d'affaires,* did not obtain his new credentials until 5 October, nearly one month later. In G. E. v. Natzmer's *Unter den Hohenzollern. Denkwürdigkeiten aus dem Leben des Gen. Oldwig von Natzmer,* 1887–9, I, 250, this dinner is placed on 11 September.

[21] Schiemann, *Geschichte Russlands,* III, 42; Martens, *Recueil,* XV, 118-119; G. Lacour-Gayet, *Talleyrand,* 1926, III, 238, believes that Talleyrand's appointment for the London Embassy made Nicholas recognize Louis Philippe, as Nicholas thought that Talleyrand would not associate himself with an unsteady régime. Reference is made to *Monsieur de Talleyrand,* a contemporary biography by the publicist C. M. Catherinet de Villemarest, I, 168, where it is said that Athalin was given a reply after Nicholas obtained a copy of the "Moniteur" which reported Talleyrand's appointment. There is no indication that this news influenced Nicholas; in any case, the reply to Athalin was given not two weeks after his arrival, as stated by de Villemarest, but another fortnight later.

new King had obtained the approval of his Imperial brother.[22]

In the latter half of September Nicholas learnt of the progress of the revolution in the Netherlands and of the various revolts in Germany. He especially sympathized with William I. The Netherlands *Chargé d'affaires* even believed that Nicholas would therefore take the peace-assurances of the French Government at their face value and help them by recognizing Louis Philippe. This he also reported to be the opinion of the British Ambassador.[23] Nicholas himself presented a different view: writing to his brother he refused to believe that warlike feelings in France would be reduced by the recognition of Louis Philippe.[24]

Accordingly, the Emperor of Russia did not conceal his real attitude towards the Orleans monarchy. Athalin did not obtain his audience of leave and in the reply which he carried with him to Paris the protocolar "monsieur mon Frère" was omitted, a fact which came quite unexpectedly to Athalin and Bourgoing, and deeply offended the King of the French.[25]

The French *Chargé d'affaires* observed with uneasiness the obvious change in the attitude of Nicholas after the Brussels and Dresden riots. This observation was confirmed by Nesselrode, who also expressed his belief in the probability of Prussian or English troops being sent to Belgium.[26]

The Russian representatives in Berlin, Vienna and London were informed that Louis Philippe was recognized after the Emperor had learned of his recognition by the other Courts. "Mais S.M. ne se dissimule point tout ce que tel encouragement, fourni à la révolution peut avoir de conséquence funeste." In spite of this unfavourable beginning the Emperor hoped that now appropriate instructions be sent by the three governments to their representatives in Paris to extract from the French Government adequate declarations regarding its intention to respect the territorial settlement of 1815.[27]

[22] Nesselrode to the Grand Duke Constantine, 29 Sept., B.P.

[23] "Les troubles survenus dans le Royaume ont vivement affecté la famille impériale. Il devait y avoir un bal à Yelagmine, ces tristes événemens y ont fait renoncer." O'Sullivan de Grass to Verstolk, 16 Sept., 22 Sept., R.A.

[24] Nicholas to Constantine, 27/8 Sept., Correspondance, p. 49.

[25] Athalin to Molé, 1 Oct., A.é. Russie; Broglie, A.L.V.C. duc de, *Souvenirs*, 1886, IV, 27; cf. Martens, *Recueil*, XV, 124-5.

[26] Bourgoing to Molé, 2 Oct., A.é. Russie.

[27] Nesselrode to Matuszewic, Alopeus and Tatistcheff, 5 Oct., B.P.

II

If Nicholas was doubtful about the choice of his own policy, he soon arrived at the conviction that he had to take decisive steps himself. On 1st October Athalin was given a favourable reply, and on the 12th and 13th respectively St. Petersburg learned of the failure of Prince Frederick of the Netherlands in his attempt to restore the King's authority in Brussels, and that Diebitsch remained unsuccessful in convincing the Prussian Government of the necessity of intervening in the Low Countries by Prussia and Russia on their own. Nicholas then resolved to mobilize his armed forces and to notify the other Powers that he felt compelled to check the revolution, and according to the Aix-la-Chapelle stipulations he offered his 60.000 men for joint action.[28]

At the same time, as he expected the King to ask the Powers of the Quadruple Alliance for armed assistance, the Netherlands *Chargé d'affaires* went to enquire with Nesselrode about the views of the Imperial Government in this respect. The Vice-Chancellor assured him that the King of the Netherlands had "non seulement le droit de compter sur l'appui de ces puissances [signatories of the Alliance] mais même celui de l'exiger." Nesselrode represented this as the opinion of the Prussian and even of the English Courts. These views of the Russian Minister were also reflected in Bourgoing's report, with the important addition, however, that the assistance in question would not be asked for, "qu'à la dernière extrémité."[29] The Dutch representative equally noted that Lord Heytesbury "paraît incliner entièrement vers la guerre et la considère non pas comme une éventualité probable mais comme un remède nécessaire" to uphold the social order in Europe. A prompt armed intervention would enable the Powers to control the situation before France could assemble her forces, the other Powers being all prepared.[31]

[28] Schiemann, Die Sendung Diebitschs, p. 18.

[29] O'Sullivan de Grass to Verstolk, 12 Oct., R.A.; Bourgoing to Molé, 13 Oct., A.é. Russie.

[30] O'Sullivan de Grass to Verstolk, 12 Oct., R.A. Yet, Heytesbury seemed somewhat uneasy when he reported that, according to Matuszewic's dispatch, "Yr. Lordship admitted the right of the King of the Netherlands to call upon the members of the Alliance to assist him in his endeavours to maintain existing engagements, and any attempt to prevent foreign interference under such circumstance would be a legitimate cause of war." (to Aberdeen, 13 Oct., F.O. Russia). A subsequent dispatch of Matuszewic, of 8 October (see Appendix I), set the matter right.

It is worth noting that at the same time both Lord Heytesbury and Jonkheer O'Sullivan de Grass reported Russian military unpreparedness. The British Ambassador describes the spreading of cholera, which rendered recruiting almost impossible, and concludes: "Russia may almost be considered as hors de combat for the moment." His colleague does not think Russia to be fit for distant foreign expeditions, now less than ever. Later, he surveys the state of the Russian and Polish armies, pointing out that both the capital and the Kingdom of Poland must not be left without appropriate garrisons, but adds that "malgré ces considérations le comte Nesselrode ne paraît pas reculer devant la possibilité d'une guerre." [31]

The official request of the King of the Netherlands for military assistance arrived in St. Petersburg on 15th October. The Emperor was eager to comply with it. "...ce qui est hors de doute", he wrote to his brother, "c'est qu'il faut se mettre sur pied de guerre". Accordingly, he advised the Grand Duke as to the choice of the officer to be left in charge of the part of the Polish army which was to be left in Warsaw. At the recommendation of Nesselrode the answer to King William's request was delayed until the arrival of news from other Courts, but Nicholas did not change his mind when he learnt of their attitude – he was determined either to make them follow his example or "himself suffer an honourable defeat". Future development he took quite seriously: "Ce n'est pas la Belgique que je crois combattre là-bas", he wrote on the margin of Nesselrode's report, "c'est la révolution générale qui de proche au proche et plus vite qu'on ne le pense, nous menace nous -mêmes, si l'on nous voit trembler devant." [32]

The official letter to William I informed him of the mobilization of troops, but declined any isolated Russian steps which Nicholas considered excluded by the very fact that the King had applied to all four Courts.[33]

Nesselrode seemed to have taken a less serious view: the intervention in the Netherlands need not mean war. In the words of the Dutch representative, Nesselrode believed "que la France, qui se

[31] O'Sullivan de Grass to Verstolk, 12 Oct., R.A.; Heytesbury to Aberdeen: "...and the Polish army... would be of a very doubtful fidelity if opposed to an army of the French." 18 Oct., F.O. Russia.

[32] Schiemann, Die Sendung Diebitschs, p. 19; Nicholas to Constantine, 18 Oct., Correspondance, p. 55; Martens, *Recueil,* VIII, 166.

[33] Nicholas to William I, 25 Oct., Ged., III, 404.

sera levée en masse contre une invasion, restera paisible lorsqu'il s'agira d'une attaque." Hence the Vice-Chancellor would welcome the landing of a British army in the Netherlands, a landing which would not be opposed by the French as they feared a rupture with England. Nesselrode knew that that could only be a small detachment, "que c'est de la Prusse que doit venir le nombre, mais que l'effet moral serait très salutaire..." This proposition was included in the instructions to Lieven and Matuszewic, together with the offer of 60.000 men by Russia.[34]

A somewhat doubtful point concerning the advance of Russian military preparations, is whether Nicholas also informed the Prussian King of his intention of embarking immediately 10.000 men for Belgium. The Netherlands Minister in Berlin regretted to inform The Hague that Frederick William III lost no time in sending a special courier to his son-in-law with an urgent request to stop this expedition. The incident, however, is reported neither by Martens nor by Schiemann.[35]

In the meantime the succession of events in the Southern Netherlands perhaps furnished some material for reflection in the Russian capital. St Petersburg was kept fully informed of the course which the Royal Government had adopted. To the Russian Minister at The Hague the situation presented a pretty grim picture. He represented the cause of the House of Orange as very precarious and did not approve of the expedition of the Prince of Orange to Antwerp, nor of the re-installation of the Minister Van Maanen, unpopular in the South, the two steps being contrary to one another. No success, he believed, could be expected from the establishment of a centre of government in Antwerp, in opposition to that of Brussels, because the dominating trends in Belgium were definitely hostile to the House of Orange. Whatever chances the Prince of Orange in Antwerp might have, were spoilt by the King, who withdrew his son's powers. Incidentally, the Prince himself wrote to his Imperial brother-in-law as if he hardly saw a way out of his situation. Gourieff, however, tried to comfort him by referring to the possibility of an armed intervention of the Powers, a possibility not excluded, he wrote, from the last instructions Talleyrand had received. This would, of course, be something entirely different

[34] O'Sullivan de Grass to Verstolk, 18 Oct., R.A.; Nesselrode to Matuszewic, 13 Oct., Martens, *Recueil*, XI, 438.

[35] de Perponcher to Verstolk, 13 Oct., R.A.

from the Prince's attempt in Antwerp, while the Russian Minister thought it wiser to remain silent about the real motive of the British Government in abstaining from such a solution, its internal weakness, although he must have been familiar with this through his correspondence with Matuszewic. Britain did not use force, he wrote, because she had proposed the conference and had to wait until negotiations reached "un certain degré de maturité." [36]

The Belgian news is reflected by Nicholas in his correspondence with the Grand Duke Constantine, and in Van Heeckeren's report of his audience with the Emperor on 14th November. Nicholas was astonished that the command of the Brussels expedition was given to a Royal Prince (i.e., Prince Frederick), and equally surprised, that the powers given to Prince of Orange confidentially were later withdrawn publicly, indeed to the astonishment of all the foreign governments. Nesselrode too pointed out the inadvisability of the restoration of Van Maanen.[37]

The autumn months of 1830 witnessed some confusion both about the views of the Court of St. Petersburg abroad, and vice-versa, caused not only by the distance between the Russian capital and those of the other countries, but chiefly by earlier opinions of Nesselrode, Pozzo and Matuszewic in contrast with those of Diebitsch. Whatever Nesselrode might have said in Berlin before his return to Russia, Diebitsch wrote on 8th September, having scarcely arrived, that the Prussian King wholly shared in the apprehensions of a general war and therefore wished him to open talks with Generals Witzleben and Krauseneck "au sujet d'une action militaire commune contre la France." Subsequently, before even the recognition of Louis Philippe by Russia was known in Berlin, it was reported that Diebitsch had moderated his views under the influence of the Prussian statesmen. Again, slightly later, the Russain Minister in Berlin informed his Court that "le gouvernement prussien était persuadé qu'une guerre européenne était inévitable." [38]

A more modest and matter-of-fact statement is delivered by the Russian Minister in The Hague: "Je crains bien que l'hésitation que

[36] Gourieff to Nesselrode, 6, 12 Oct., Prince of Orange to Nicholas, 17 Oct., Ged., III, 400-404; Gourieff to Prince of Orange, 25 Oct., F. Gerretson, *Muiterij en scheuring*, 1936, II, 264.

[37] Nicholas to Constantine, 9 Nov., Correspondance, p. 60; Van Heeckeren to Verstolk, 14 Nov., R.A.

[38] Diebitsch to Nicholas, 8 Sept., Alopeus to Nicholas, 10 Oct., Martens, *Recueil*, VIII, 166; Luxburg to the King of Bavaria, 3 Oct., de Guichen, p. 162.

les puissances témoignent à intervenir dans la Belgique ne leur devienne funeste. Elles veulent éviter la guerre et délibèrent, la France se fortifie de progrès de la propagande." This is Britain's fault as London is influenced by Talleyrand. "Si dans le premier moment l'Angleterre s'était déclarée avec force, je ne puis pas en douter que la France se serait résignée à l'intervention aussi dans les affaires de la Belgique." Unlike on later occasions, there is here a marked coincidence of views with those of St. Petersburg itself.[39]

This situation found an echo in the observations made by the Western diplomats as well. Early in October Talleyrand believed that, Prussia and Austria having made no promises of armed assistance to the King of the Netherlands, it was most likely that Russia would follow the same course. News from England, where Charles X was reported to have obtained an unlimited credit with an Edinburgh banker which was attributed to the support of the Russian Emperor, soon made the French Ambassador consider the situation in less bright colours. Not only the Emperor of Russia "ne voit que l'emploi de la force pour arrêter l'envahissement des doctrines de désordre", but there is also a war party in Prussia, observed the Ambassador. "C'est une époque critique que les événements de Belgique ont bien compliquée." [40]

Talleyrand believed that Metternich was for peace, but this was not exactly the case. Much in the same way as Nicholas, the Austrian Chancellor deplored the lack of a uniform attitude of the Powers towards Louis Philippe. When at the end of August he learned of the Russian measures against French subjects, he was very much surprised as they did not conform to the letter of the "chiffon"; but he followed, though with some hesitation, the example given by Prussia and approved by Nesselrode. When, however, it became known that Nicholas was against the recognition of the Orleans King, he "again lamented that Count Nesselrode had not consented to suspend all measures relating to France until the great Continental Powers determined upon some definite plan for their guidance." Nevertheless, both the Chancellor and the Austrian Ambassador

[39] Gourieff to Nesselrode, 19 Oct., Ged., III, 410.

[40] Talleyrand to Molé, 8, 11, 29 Oct., G. Pallain, *Ambassade de Talleyrand à Londres*, 1891, pp. 20, 25, 44. Quoted as Pallain. The Netherlands Ambassador Falck noted as early as 6 October (to Verstolk, Falck Papers, R.A.) that Matuszewic kept silent for want of instructions. He received them in Nesselrode's dispatch of 13 October, which arrived in London after Talleyrand wrote the last of his dispatches quoted here.

in Paris adopted an uncompromising attitude with regard to the French Government.[41]

III

In the meantime there was a new development in Belgian affairs. The provisional authorities in Brussels declared Belgium independent and summoned a congress to work out a frame of government for the new state. In London on 4th November a conference of the representatives of the five Powers assembled with a view to a settlement for the Netherlands, an armistice being considered to be the first task. The Russian *Chargé d'affaires,* Matuszewic, took part in this meeting, a step later approved by the Imperial Cabinet.[42] But Gourieff in The Hague observed disapprovingly that, apart from military considerations, the proposal of adopting the old frontier line between the Dutch Republic and the Austrian Netherlands as a demarcation line for the armistice, embodied by the Conference in its first protocol, might easily create the impression as if it concerned two distinct political units.[43]

The Russian government was far from accepting this view. They regretted that the English government renounced – at least for the time being – the idea of armed intervention and proposed a conference to which France was also to be a party. The Vice-Chancellor's dispatch of 31st October surveys the difficulties connected with this project, stressing: "que cette négociation, loin de décider immédiatement la question principale, celle de la soumission des insurgés de la Belgique, ne fait que l'ajourner; qu'elle absorbera un tems précieux qu'il aura été essentiel d'employer efficacement; qu'elle mettra les rebelles à même de consolider leur coupable entreprise, fournira un funeste encouragement aux révolutionnaires des autres pays et ajoutera ainsi aux difficultés et obstacles que nous aurons à combattre lorsqu'en dernière analyse il sera d'un commun accord jugé indispensable de faire intervenir la force des armes." [44]

[41] F. von Gentz, *Tagebücher,* 1920, p. 186; Cowley to Aberdeen, 9 Sept., to Palmerston, 19 Dec., F.O. Austria. The latter dispatch, from which the quotation is taken, surveys Austrian policy with regard to France from the beginning of the July revolution.

[42] Nesselrode to Lieven, 30 Nov., Ged., III, 426.

[43] Gourieff to Nesselrode, 11 Nov., Ged., III, 421.

[44] Nesselrode to Matuszewic, 31 Oct., Ged., III, 412, reprinted from the *Mémoires* of Talleyrand, III, 363. The editor of the *Gedenkstukken* corrected the apparent error of Talleyrand (or de Broglie's), who had repeated the original date, 19 October (old style).

Nevertheless, for the sake of unity among the Powers, the Emperor acceded to the idea of the conference, its aim being described as "la pacification du Royaume des Pays Bas, moyennant un changement dans les conditions de l'union de la Belgique avec la Hollande, mais avec le maintien de l'intégrité de cet Etat sous la domination de la maison d'Orange." The settlement arrived at would be supported by the four Powers, and "quelle que soit d'ailleurs l'opposition de la France, elles emploieront même la force des armes, s'il était nécessaire, pour mettre l'engagement convenu à l'exécution." The instructions for Matuszewic mention further the objections which may be raised by France and which should be met with an adequate answer. Lastly, the Russian Government feels that, if Britain "se voit hors d'agir à présent," it is essential that she use the time devoted to the negotiations for improving her armed forces.[45]

For the rest, it seemed that the Imperial Cabinet hoped that Britain might, if she would not join the intervention, at least not oppose it.[46]

To Gourieff Nesselrode put the case even more clearly. The Emperor is prepared to offer to the King of the Netherlands military assistance. "Pour sa part, l'Empereur est tellement pénétré de cette conviction, qu'il n'a point hésité ... même à ordonner dès à présent, ainsi qu'il en informe S.M. [Néerlandaise], la réunion d'une armée, qui sera prête de franchir les frontières de l'Empire dès que sa coopération sera jugée nécessaire." This step is, however, delayed by the conference now to be held, while the King of the Netherlands himself appears to be trying other means to appease his Southern subjects. Both measures, the project of the Lieutenant-Generalship of the Prince of Orange in Belgium as well as the Conference itself "n'offrent malheureusement que peu d'espoir de succès." Nevertheless Nicholas would act in common with the other Powers; only the "étrange prétention du gouvernement français" to oppose all armed intervention would never be admitted.[47]

[45] Same dispatch.

[46] Matuszewic to Nesselrode, 8 Oct.: "Si la France veut soutenir maintenant le principe qui exclurait à l'avenir toute intervention militaire, toute mesure coërcitive, l'intention du cabinet de St. James est, non de souscrire à cette demande, ou de laisser croire que jamais il y souscrira, mais d'ajourner la discussion, en la déclarant prématurée." This is an extract from Matuszewic's dispatch, appearing in Appendix I. The copy of this dispatch was appended to Van Heeckeren's dispatch of 13 November, R.A. It is also translated into English, word by word, in Lord Heytesbury's dispatch of 22 October, F.O. Russia, further evidence of the importance attached to it by the Russian Cabinet.

[47] Nesselrode to Gourieff, 31 Oct., Ged., III, 415-6.

Bourgoing as usual pictured the situation in brighter colours. The Russian Government welcomed the Conference because, as Nesselrode had told him, St. Petersburg being too far away from the theatre of events, no proper judgment about their development and consequences could be formed. Yet, according to the Vice-Chancellor, the main problem to be solved by the conference was: the maintenance of the rights of the House of Nassau, as well as the maintenance of the political connection between the two parts of the Kingdom of the Netherlands. The possibility of an armed intervention, regretted the French *Chargé d'affaires*, however distant, "n'est point, toutefois, perdue de vue." [48]

On 8th November the Netherlands Minister, Baron van Heeckeren, to whom Bourgoing ascribed considerable influence with the Russian Court, returned to St. Petersburg after two months spent in Holland, and two days later went to see Nesselrode.[49]

The Vice-Chancellor began his interview by pointing out that in his opinion only an armed intervention could restore order in the Belgian provinces, but the course of events rendered the isolated action of one Power impossible. He nevertheless believed that the Conference should open by declaring that its members were prepared "à faire respecter leur décision par la force des armes." If France were to oppose such a declaration or measures undertaken as its result, it could only be considered as a declaration of war. Russia would take that risk, and the Imperial Government believed in the concurrence of its allies, including England, as all reports from London confirmed Wellington's intention of keeping to the letter of the treaties.[50]

In his two following dispatches the Netherlands Minister tried to give an appreciation of Russian preparedness. Nicholas had declared to him that all decisions taken in London should be supported by military force; he added that he had given evidence of this determination by ordering that publicity be given to the announcement of the Russian mobilization.[51] The Emperor further believed that all his allies, Wellington, "ce plus timide de tous", included, shared this opinion. "Si nous avions partout des amis aussi chauds

[48] Bourgoing to Molé, 30 Oct., A.é. Russie.

[49] Bourgoing to Molé, 9 Nov., 7 Dec., A.é. Russie; cf. J. de Bosch Kemper, *Geschiedenis van Nederland na 1830,* 1873, I, Letterkundige Aantekeningen, p. 111.

[50] Van Heeckeren to Verstolk, 10 Nov., R.A.

[51] The Imperial ukase announcing the Russian mobilization was published in "Journal de St. Pétersbourg" of 15 November.

que les Russes", observed the Dutch Minister, "nos affaires seraient bien plus avancées." [52] He did not, however, omit to report less favourable circumstances. He gave a lengthy description of an incident between Polish and Russian officers in a Warsaw theatre, with the anxious comment that even the smallest spark was likely to cause a blaze. Fortunately, van Heeckeren thought, the Grand Duke Constantine caused all the officers concerned, Polish and Russian, to be arrested.[53]

Bourgoing at this time considered the military preparations all over Europe. "*Les affaires de France en elles-mêmes n'ont jamais donné à personne l'idée d'une intervention armée,*" he wrote, "mais à plusieurs reprises la tournure que prenaient les choses en Belgique en a fait aborder la possibilité". This was particularly true with regard to Emperor Nicholas. The French *Chargé d'affaires* reviewed the increasing difficulties which Nicholas would experience in trying to assemble an army of 150.000 to 200.000 men, which he had promised as the Russian contingent in case of a general war. Yet, "pour qui connaît le caractère personnel de l'Empereur, il ne peut s'élever aucun doute sur la certitude de le voir persévérer dans ce qu'il a promis de faire, quelque grand que puisse devenir le sacrifice." [54]

It should be added to these remarks that Nicholas was not unaware of the cautious policy of the Powers who were his allies in a possible conflict with France, and of the attitude of Frederick William III in particular. The King of Prussia could very easily help his cousin in the Netherlands by increasing his forces in Luxemburg (the capital of which, as a Federal fortress, always possessed a small Prussian detachment), on either King William's or the Federal German Diet's request, without being obliged to ask the consent of the other Powers. But the Berlin Government, aware of the disapproval of Russian re-armament not only by France but also by Britain, thought it wise to refuse any measures which might be considered warlike; this was regretted by Nicholas who thought that even the occupation of Maastricht, in agreement with The Hague, would not require the consent of the other Powers. [55]

The embarrassing mission of Diebitsch in Berlin continued and its

[52] Van Heeckeren to Verstolk, 17 Nov., R.A.

[53] Van Heeckeren to Verstolk, 13 Nov., R.A., one of the few marked "Top secret".

[54] Bourgoing to Molé, 9 Nov., A.é. Russie.

[55] W. Gronemann, *Die Haltung Preussens in der Belgischen Frage,* 1928, p. 8. For Nicholas' opinion cf. Martens, *Recueil,* VIII, 166

echoes made their way to Paris and London. There was no doubt in Paris that the Russian envoy was negotiating some decisive step with regard to France, and though it was believed that he would be influenced by the more moderate attitude of the Prussian Government, the news of re-armament in Russia could not fail to alarm France. Firstly, Prussian good offices were to be employed to get to know, if the Russian armaments "annoncent de projets sérieux de guerre ou seulement intention de prendre une attitude militaire propre à donner à la Russie plus d'influence dans les affaires des Pays Bas." The French representative in the Russian capital was also instructed to get information on the subject but carefully to avoid creating the impression of demanding explanations "plutôt qu'un désir bien naturel de s'éclairer sur des faits qui doivent attirer si vivement notre attention." This was not all that troubled the French Government; Talleyrand also observed the increasing danger to peace in the intrigues of Metternich, whereas Diebitsch had dealings of his own with the Prussian war party.[56]

Eventually, the direction of Russian policy seemed irrevocably established when the official news of the Russian mobilization spread in the West. By that time, the end of November, 1830, the French Government had already made various efforts to calm down national exasperation in Belgium. Louis Philippe may have realized that the revolution in Belgium undermined his throne, yet he could not prevent the French supporting the Belgians in every possible way.[57] This was not the unanimous opinion of the contemporaries: there were diplomats in Paris who maintained that the French Government would be able to control its unruly subjects in case of a British landing in the Southern Netherlands, a landing which could have a salutary effect on the developments there.[58] The British Government, however, was not prepared to take this course. Further proof, therefore, was required from the French Government, if it wanted its neutrality to be believed, while in the meantime the Powers would

[56] Maison to Mortier, 19 Nov., to Bourgoing, 22 Nov., A.é. Prusse, A.é. Russie; Talleyrand to Maison, 20 Nov., Pallain, p. 86.

[57] For the part taken by the French agents in the early stages of the Belgian revolution see R.E. Dinger Hattink, *De Brusselsche opstand,* 1930. The opposing opinions on this controversial question are once more reviewed by Professor Demoulin in "L'influence française sur la naissance de l'Etat belge", *Revue historique,* CCXXIII (1960), p. 13.

[58] Fagel to Verstolk, 4, 8 Oct., Gerretson, II, 364; K. Hillebrand, *Geschichte Frankreichs,* 1878, notes on pp. 144-6.

behave as if France was not neutral. This was the meaning of the answer given by Nesselrode to Bourgoing on the latter's enquiring what was the purpose of the Russian mobilization. The Vice-Chancellor turned the question to the corresponding movements in France and elsewhere, but, moreover, to the troublesome presence in Belgium "d'une multitude de Français, publiquement recrutés à Paris." A public disavowal of this proceeding by the Government was required, Nesselrode said.[59]

Before this report reached Paris, further concessions were made by the French Government for the sake of foreign relations. Thus Marshal Maison, only just installed in the Foreign Ministry, was replaced by Count Sebastiani, and the new Minister made a special effort to prevent the vote of exclusion of the House of Orange by the Belgian Congress.[60]

So far the French Government could congratulate itself: the establishment of the London Conference put off the danger of immediate intervention in Belgium. But this did not necessarily mean the dissolution of the union between the Southern and Northern Netherlands, as loudly claimed by public opinion in France; nor did the Conference in its first protocol make any provision for special treatment of Bouillon and Marienbourg, two places lost by France after the "Hundred Days", whose reunion with France was equally part of the current public demand. The occupation of these places, incorporated by King William in the Grand Duchy of Luxemburg, whether effected by the Prussian troops, as proposed by the German Federal Diet, or by others would produce the worst of impressions in France, whilst moreover "les députés du duché de Luxembourg qui assistent au congrès [de Belgique] sont tous des orangistes zélés", as Sebastiani observed.

All these matters were present in his mind, when he regretfully stated that in the present circumstances, the French Government was yet bound to support the candidature of the Prince of Orange for the Belgian throne. He surveyed the state of affairs in Belgium and

[59] Van Heeckeren to Verstolk, 10 Nov., R.A.; Bourgoing to Molé, 11 Nov., A.é. Russie.

[60] Molé to de Barante, 17 Nov.: "Le Maréchal Maison est outré et on illumine à l'Ambassade de France à Londres et à celle de Londres à Paris. Quant à la Russie et son ambassadeur, il faut chercher la mesure de leur satisfaction dans les colonnes des "Moniteurs" renfermant la politique du général Sebastiani." P. de Barante, *Souvenirs*, 1890–1901, IV, 15. The rôle of Maison in the Guilleminot affair will be shown in Chapters III and IV, see below, p. 98 and 142.

concluded that there was a possibility of the Prince's success, whereas there was none for putting forward the candidature of a French prince. Before, however, Sebastiani was able to close his dispatch to Talleyrand he received word that the Belgian Congress was inclined to pronounce the vote of exclusion of the House of Nassau.[61]

This the French Government could not allow to happen: the other Powers were sure to intervene in such a case. A special envoy, Langsdorff, was sent off immediately to Brussels with instructions to make the Belgians understand "que cette déclaration... serait une mesure imprudente, impolitique et tellement grave que leur indépendance même s'en trouverait compromise.... La guerre serait la suite presque inévitable d'un acte aussi décisive et la France ne veut pas y être entraînée malgré elle." [62]

Langsdorff's mission did not prevail: on 24th November the vote of exclusion was passed in Brussels. But the effort of the French Government had not been made in vain, Talleyrand thought; the other Powers did notice it. The Ambassador's views were, however, not optimistic: a few days earlier he observed "L'Europe est certainement, en ce moment, dans un état de crise. Eh bien, l'Angleterre est la seule puissance, qui, comme nous, veuille franchement la paix." [63]

Sebastiani had read this opinion when he set himself the task of appeasing St. Petersburg in a squeer way, by means of his instructions to Bourgoing, dated 1st December. Langsdorff's mission was first to be discounted. The vote of the Belgian Congress is qualified as "une décision fâcheuse", which "d'autant plus [a] affligé le Roi qu'il s'était efforcé de la prévenir et qu'il n'a cessé de faire des voeux sincères pour l'élection d'un prince de la famille de Nassau." The main theme of the dispatch is, however, the Russian mobilization, which is commented on in a tone varying from apologetic reproof to astonishment. The re-armament of France's neighbours might be explained by their anxiety with regard to developments in France, but this argument would not be valid for Russia, safe through her distant situation, unless the assumption was made that a new anti-French coalition was being created. The French Government hoped that Emperor Nicholas would save Europe from the horrors of a new war; similarly France would do all "ce qui sera compatible avec l'honneur et la dignité de la France" for the same purpose.

[61] Sebastiani to Talleyrand, 22 Nov., Ged., II, 104.
[62] "Instructions à MM. de Langsdorff et Bresson", Ged., II, 91.
[63] Talleyrand to Sebastiani, 30 Nov., Ged., II, 104 and 27 Nov., Pallain, p. 99.

If the above reasoning differed form what used to be the official French attitude towards Russia, the remainder of this dispatch was nothing else but a threat:

"Nous aimons, d'ailleurs, nous en rapporter à l'assurance, qui nous a été exprimée par M. de Nesselrode, que les armemens de la Russie n'annonçaient aucun projet d'attaquer la France. Mais de même qu'il vous a dit que ces armemens ne seraient destinés contre nous que du moment où les armées françaises dépasseraient nos frontières, vous voudrez bien, Monsieur, lui déclarer aussi, que, si les armées russes quitteront leur territoire pour entrer en Prusse ou dans tout autre Etat de l'Allemagne, la France alors se verrait forcée de regarder l'état de paix comme rompu et n'aurait plus de prendre conseil que de son honneur et ses intérêts."

The comment with which the French *Chargé d'affaires* is instructed to supplement his momentous declaration adds but little, if anything, in the way of softening it:

"...ce ne sont nullement des menaces que nous voulons faire entendre, qu'un semblable langage nous est naturellement inspiré par un caractère de franchise qui ne se démentira jamais par un ferme désir de prévenir tout ce qui pourrait troubler ou compromettre la paix. La politique du Roi est toute de vérité et de bonne foi, et l'on peut être fermement persuadé en Russie que la tranquillité générale est le plus cher de nos voeux." [64]

It is difficult to assume that this step of the French Government could have passed in St. Petersburg without causing a major upheaval, and, indeed, perhaps sealing the Emperor's determination. As it happened, the dispatch arrived in the Russian capital when circumstances had entirely changed.

Towards the end of November Nesselrode seemed satisfied with the course of events around the Belgian question. He spoke approvingly of the first pronouncements of the French cabinet which took office on 2nd November. The opening of the British Parliament gave rise to the following comment: "Si le duc de Wellington se maintient au Ministère, on ne doute pas que l'Angleterre ne consente à intervenir à main armée, si les révoltés de Belgique refusent de se soumettre aux décisions des conférences de Londres." This statement carried some weight as in his instructions to Lieven (at that time en route to London) the Vice-Chancellor expressed his belief

[64] Sebastiani to Bourgoing, 1 Dec., A.é. Russie.

that Wellington's government, "une fois engagé dans une intervention collective" – which term probably denoted the London Conference only – could not, "malgré les doctrines professées en Angleterre", refuse his share in all measures which the pacification of Belgium would render indispensable. Unfortunately there was a possibility of Grey's entering the Government, in which case armed intervention was almost impossible. This opinion was written after Grey had actually taken office.[65]

The Imperial Cabinet maintained their views on the solution of the Belgian problem. In view of the information that the French Government inclined towards the complete independence of Belgium, with a separate crown, Nesselrode told the Netherlands Minister that new instructions were being sent to Lieven which distinctly forbade him to consent to such an arrangement. These instructions indeed were clear-cut: "la seule combinaison admissible... serait que la Belgique eût un gouvernement séparé de celui des provinces Hollandaises, mais soumis au même sceptre; que le Roi continuât à rester souverain des deux pays, et que l'héritier de la couronne devient son lieutenant dans les provinces du midi." [66]

On this subject the French *Chargé d'affaires* had a difficult interview with Nesselrode. The Vice-Chancellor continually returned to the idea of supporting the Prince in the Southern provinces by force of arms. Yet Bourgoing found the feeling in St. Petersburg less hostile to France. The French reply to the Russian projects for a settlement in the Netherlands was well received. Moreover, Nesselrode himself gave evidence of an accommodating attitude by requesting Bourgoing to suppress the sentence "La France ne renoncera qu'en désespoir de cause à ce moyen [i.e., the lieutenant-generalcy of the Prince of Orange in Belgium] de conciliation" in the copy of Sebastiani's dispatch which was to be presented to Nicholas. On the other hand, the spirit of Bourgoing's instructions remained unacceptable in St. Petersburg: the Vice-Chancellor pressed Bourgoing to name an alternative to force for making the Belgians listen to projects unanimously approved by the conference. From the conversation on this subject which ensued, the French *Chargé d'affaires* drew the conclusion "que cette question d'un premier emploi de la

[65] Van Heeckeren to Verstolk, 26 Nov., R.A.; Nesselrode to Lieven, 30 Nov., Ged., III, 426-7.

[66] Van Heeckeren to Verstolk, 29 Nov., R.A.; Nesselrode to Lieven, Ged., III, 426-7.

force, sans être entièrement abandonnée, sera du moins reléguée parmi les partis extrêmes auxquels on ne voudra recourir qu'après longue et mûre réflexion." [67]

In reality the situation seemed much more serious and the moment was approaching when military action would not be merely a menace. On 18th and 21st November respectively the Minister for War – for the Russian troops – and the Grand Duke Constantine for the troops of the Kingdom of Poland and the Western provinces of Russia, formerly belonging to Poland – reported to Diebitsch, who still stayed in Berlin, that their armies would be ready to march by 1st January, 1831. Earlier the Secretary of State for the Kingdom of Poland informed the Minister of Finance in Warsaw of the Emperor's satisfaction with the progress of the military and other measures connected with the war. No marching orders, however, had been issued to the troops; this would be done after the return of Diebitsch. "Il a ordre de se diriger à son retour de Berlin par Varsovie pour se concerter avec S.A.I. Mgr. le Grand Duc Césarevitch sur tout ce qui concerne le mouvement des armées." [68]

The Prussian king preferred to expose himself to accusations of sharing Diebitsch's warlike opinions rather than to let him return to Russia. In spite of the Field-Marshal's renewed requests Frederick William kept him in Berlin. The execution of the Russian plans could not but end in war: if credit is to be given to Bourgoing's opinion of Nicholas, the Emperor would have been able to overcome the opposition of the majority of his advisers who were against the war.

But the issue did not materialize. On 3rd December Berlin learnt of the outbreak of a military rising in Warsaw. Diebitsch, no longer wanted, hurriedly left for St. Petersburg via the shorter route through East Prussia.[69]

[67] Bourgoing to Sebastiani, 7 Dec., A.é. Russie.

[68] Schiemann, Die Sendung Diebitschs, p. 21; Grabowski to Lubecki, 17 Nov., *Korespondencya Lubeckiego,* (The Correspondence of Lubecki) ed. St. Smolka, Cracow 1909, III, 397.

[69] Schiemann, Die Sendung Diebitschs, p. 21. For Bourgoing's opinion see above, p. 50.

CHAPTER II

THE OUTBREAK OF THE REVOLUTION IN POLAND AND THE INDEPENDENCE OF BELGIUM

It has been observed earlier that the Netherlands Minister in St. Petersburg, impressed no doubt by the state of affairs in other parts of Europe, attributed some importance to a mere incident between the Russian and Polish officers in a Warsaw theatre. A more serious matter was reported from Berlin in the latter half of November: a military plot had been discovered in Warsaw, followed, the report stated, by a most rigorous inquiry. The Netherlands Minister in Berlin, Count de Perponcher, commented on this report as follows: "Je dois *à cette occasion* porter à la connaissance de V.E. que des agens du comité directeur de Paris parcourent l'Allemagne." Besides, doubts as to the fidelity of the Polish troops in a fight against the French had been expressed by the British Ambassador in Russia already in October.[1]

But the Warsaw riots came unexpectedly even to the higher circles in Poland. During several weeks the Warsaw events must necessarily, to foreign observers, have left uncertainty as to their ultimate course and, even after the Polish Diet had declared on 20th December that the outbreak of 29th November had initiated a national movement, the possibility of a settlement without recourse to arms still existed; envoys, sent from Warsaw to the Russian capital, did not come back until January.

The impact of the revolution in Poland upon the state of affairs elsewhere – and especially upon the London Conference on Belgian affairs – was primarily caused by the very first news of its outbreak. The Conference had already achieved its first object in establishing a suspension of hostilities between the Royal forces and the insurgents, but complaints were continually made by both sides as to its observance, and the Scheldt was kept closed by the Dutch navy, much to the annoyance of British and other merchants. The

[1] de Perponcher to Verstolk, 24 Nov., R.A. Italics mine; cf. above, p. 43 and 50. The inquiry in Warsaw was, however, far from rigorous, as the Grand Duke Constantine, C.-in-C. of the Polish army, disliked any investigations of *his* men.

riots which took place in Warsaw on the evening of 29th November were known in Berlin on 3rd December, in Vienna on the 6th, and in London and Paris on the 10th, while Nicholas learnt about them late at night on the 7th; on 8th December the news became public in the Russian capital. It must be borne in mind, therefore, that, when Talleyrand began his campaign in the Conference for the recognition of the independence of Belgium, he knew that the Russian Ambassador, aware of the change in the political configuration brought about by the Polish news, would not be able to ascertain his Imperial master's reaction to it.

Far from bringing about a concrete result, the Protocol of 20th December created a series of new problems, foremost that of the frontiers of the new state, in which the special position of the Grand Duchy of Luxemburg had to be considered, while the second important question demanding an immediate solution was that of the Belgian Crown.

I

Even before the Warsaw riots were known in St. Peterburg, this news arrived in Berlin and Vienna. Frederick William III received it as he was writing to his daughter, the Empress Nicholas, and could not find words strong enough to describe his anxiety. But the British Minister at Berlin, Chad, reporting the event two days later, stressed the point that the proclamation issued by the authorities the day after the riots was made in the name of Nicholas, Emperor of Russia and King of Poland; Chad considered this to be a promising circumstance.[2] Again, the Netherlands Minister in Vienna qualified the movement as "émeute"; this was positively not the view of Metternich as imparted to the Prussian representative. The Austrian Chancellor called it the most disastrous blow which could have been struck at the three Eastern Powers, and concluded that Belgium was now to be left to her own fate.[3]

[2] Frederick William III to Empress Nicholas, 3 Dec., *Aus dem letzten Jahrzehnte Friedrich Wilhelm III.* Briefe des Königs an seine Tochter Charlotte, Kaiserin von Russland, ed. P. Bailleu, Hohenzollern Jahrbuch, 20, 1916, p. 153. The divided opinions in the King's immediate entourage are reflected in *Denkwürdigkeiten aus dem Leben von Leopold* von Gerlach, 1892–3, I, 59; Chad to Aberdeen, 5 Dec., F.O. Prussia. Words: "and King of Poland" underlined in the text of the dispatch.

[3] Van Spaen to Verstolk, 6 Dec., R.A.; Maltzahn to Bernstorff, 7 Dec.,

Within the next few days after the first news diplomatic reports from the three Eastern capitals were filled with reflections on the state of affairs in the light of events in Poland.

A subsequent report from Berlin said that in the Prussian capital the rising in Poland was considered as the result of the announced mobilization of the Polish forces and of the probability of their being employed to fight the French. Chad even believed that, on his departure, Diebitsch obtained a personal message for Nicholas from the Prussian King amounting to a statement that sovereigns themselves were the cause of revolts. In any case, the British Minister was told by Ancillon, the Under-Secretary for Foreign Affairs, that the warlike speech of French Prime Minister Laffitte was the result of the military preparations ordered by Nicholas. Ancillon hastened to add that the dispatch from Sebastiani to the French mission in Berlin was "of the most satisfactory nature." Meanwhile, he somewhat surprisingly continued, the Russian measures "had no object [in view]. The Emperor does not wish for war, nor can Russia act without Prussia and Austria – we all desire peace, and the French Government is beyond all doubts sincere in its professions on that subject; yet if every Power adopts precautionary measures those measures may come in contact with each other and themselves produce war." [4]

This unexpected declaration is, however, counterbalanced by a contrary statement made to Mortier, the French *Chargé d'affaires*. Ancillon said that the Russian military preparations had been justified and the Polish revolution furnished evidence to this effect. Mortier noted Prussian military measures with regard to Poland: 60.000 men had been sent to Prussian Poland, to which force, if necessary, another 30.000 would be added. In conclusion, the French *Chargé d'affaires* wrote that Ancillon and other Prussian observers admitted that France was going to exercise a much stronger influence in Belgian affairs, and that on her now depended the peace of Europe. Similar

---

K. Hillebrand, *Geschichte Frankreichs,* I, 158. The relative, characteristic, passage of Hillebrand runs: "Am morgen des 30. November war Rusland vom übrigen Europa durch einen gewältigen Militäraufstand tatsachlich getrennt; Preussen genötigt seine Aufmerksamkeit und seine Mittel zwischen die West- und Ostgrenze zu theilen, Oesterreich vereinzelt und durch die Furcht der Ansteckung seiner ausserdeutschen Besitzungen gelähmt. 'Das ist der verhängnisvollste Schlag, der den drei Hofen beigebracht werden konnte', rief der Staatskanzler, 'mann wird nun Belgien seinem Schicksal überlassen müssen.'"

[4] Chad to Palmerston, 6, 10, 12 Dec., F.O. Prussia.

opinions were attributed by Mortier to other members of the diplomatic corps in Berlin.[5]

In Vienna the French Ambassador found that the Austrian Government had known as long as two months before that a violent movement was to be expected in Poland. Consequently, the Chancellor dismissed with a mere "mais la Pologne!" any discussion on Russian re-armament. For the Dutch observer there would be "incalculables conséquences... de cette sanglante catastrophe" in Poland: the whole of Germany and most other countries were in an excited state. France, then "résistera-t-elle à la séduisante tentation d'un agrandissement, si vivement désiré, même si hautement réclamé et qui semble à présent lui tomber à soi même dans les bras?" It seems that the feeling prevailing in Vienna did not produce only this opinion; if Metternich was convinced that all revolutions originated from France, the influential Gentz saw in the Polish events, above all, a check to the great danger of the Russian plans for intervention.[6]

Of the three convinced supporters of the King of the Netherlands the warmest and at the same time the one most affected by the Polish insurrection, was the last to learn of it. Nicholas appeared calm while himself announcing the news to the troops on the morning of 8th December, and, satisfied with the spirit shown by his soldiers, seemed confident to the members of the diplomatic corps, who surrounded him during the levée, of a speedy crushing of the revolt.[7] His primary concern was for the safety of his brother from whom he had had no news for several days. His great anxiety is, however, reflected in his letter to his sister, the Princess of Orange, in which he appeared unable to say anything about the support which the Royal family eagerly expected from him. "Voilà donc que la révolution et tout ce qui s'en suit nous a atteints aussi comme je le prévoyais", he wrote. He soon received proof that he had not been wrong about the origin of all revolutionary troubles: the King of Prussia sent him an intercepted letter from a Polish leader, Jelski, to the French Premier Laffitte. Writing to his brother, Nicholas

[5] Mortier to Sebastiani, 14 Dec., A.é. Prusse.

[6] Maison to Sebastiani, 12 Dec., A.é. Autriche; Van Spaen to Verstolk, 13 Dec., R.A.; A. Prokesch von Osten, *Aus den Tagebüchern,* Vienna 1909, pp. 68-9; Gentz to Prokesch von Osten, 8 Dec., *Aus dem Nachlasse des Grafen Prokesch-Osten.* Briefwechsel mit Herrn von Gentz u. Fürsten Metternich, Vienna 1881, I, 408.

[7] Van Heeckeren to Verstolk, 8 Dec., R.A.; Bourgoing to Sebastiani, 8 Dec., A.é. Russie.

quoted Jelski's words: "C'est pour la seconde fois que la Pologne suit l'exemple de la France; en sera-t-elle la seconde fois la victime?"; "C'est clair et positif", commented the Emperor.[8]

In the meantime, the Polish rising bore its first fruits in Russian diplomacy. In his first interview with the Vice-Chancellor after the event the French *Chargé d'affaires* expressed his regret to hear of it, and his hope that a peaceful settlement might be achieved; Nesselrode, however, doubted the possibility of such a solution; all measures taken at St. Petersburg, and especially the Imperial proclamation to the Poles, indicated that Nicholas was resolved only to accept unconditional submission. But only two days after the interview Bourgoing was shown a dispatch to be sent to Pozzo di Borgo with regard to Polish affairs. It recalled the French assertion that armed intervention in Belgium was the most certain way to war as the French nation would not allow foreign armies to operate against a kindred people; but in their anxious desire to avoid war the French Government had been exerting themselves to keep public opinion quiet. The Imperial Government, therefore, urged France to show "la même loyauté dans la question de Pologne qu'elle l'a faite dans celle de la Belgique." When showing this dispatch, Nesselrode added a verbal declaration to the effect that the best way of gaining the confidence of Russia was "by abstaining from all interference, whether open or secret, in Polish affairs." [9]

The French *Chargé d'affaires* confided both the contents of the dispatch and the words of the Vice-Chancellor to Baron van Heeckeren who passed them on to the British Ambassador. The two diplomats had nothing but severe criticism for the Russian *démarche*. "I cannot think this declaration very prudent or very likely to produce the desired effect at Paris", wrote Heytesbury. "It savours too much of alarm." [10] The Dutch comment was even more outspoken. The Russian démarche "n'empêchera rien si la France est mal intentionnée, et dans le cas contraire, elle donne à celle-ci beau jeu pour faire valoir sa neutralité comme un immense service, rendu à la Russie, et pour exiger plus tard qu'elle s'en montre reconnaissante." In any case it constituted a proof of great weakness with regard to

[8] Nicholas to Princess of Orange, 16 Dec., Lie MSS (See Appendix II); Nicholas to Constantine, 20 Dec., Correspondance, p. 68.

[9] Bourgoing to Sebastiani, 8 Dec., A.é. Russie; Van Heeckeren to Verstolk, 13 Dec., R.A.; Heytesbury to Palmerston, 13 Dec., F.O. Russia.

[10] Heytesbury to Palmerston, 13 Dec., F.O. Russia.

France, an attitude of which the Russian Government had been extremely critical, where other governments were concerned. Incidentally, Bourgoing himself, who for the rest was convinced of the superiority of the Russian forces in an approaching contest, was very modest in recounting this occasion.[11]

A few days later new instructions were sent to Lieven. A survey of Russian policy so far, according to the principle: "Conserver ces traités [of 1815] et défendre l'état de possession qui en résulte" is followed by this commentary on the Polish revolution: "Cet événement, quelque grave qu'il soit en lui même", the dispatch runs, "*...ne change point l'attitude que S.M. a prise dès origine, à l'égard des affaires générales de l'Europe.*" On the contrary, the defeat of the rebels "servira à décourager d'autres contrées des tentatives criminelles" and thus would contribute to the cause of peace. Russia's principles as to Belgian policy still remain: "1. de conserver la possession de la Belgique à la maison de Nassau. 2. de consentir, en revanche, à une séparation complète que les Belges et les Hollandais réclament avec une égale insistance, pourvu que la demande en soit formellement exprimée par le Roi des Pays Bas 3. *de ne point admettre une combinaison qui placerait la Belgique sous la domination d'une dynastie étrangère* à celle qui l'a gouvernée pendant un règne de quinze ans, sous la garantie des traités." Were the Conference to proceed to abandon the rights of the House of Nassau, the Russian representatives should refer to the Emperor who would then concert with other Courts on ultimate measures. Then follows a disquieting passage: "Il serait prématuré d'en tracer la ligne, aussi longtemps que nous ne connaissons pas les voeux que S.M. Néerlandaise serait dans le cas d'émettre pour le maintien de ses droits, *ni les sacrifices qu'elle pourrait préférer peut-être de porter à la tranquillité générale.*" In the meantime, "nous avons l'espoir certain de rétablir en Pologne le calme et la sécurité et de combler ainsi l'abîme que les révolutionnaires ouvraient sous nos pas."[12]

A copy of the instructions to Lieven was sent to Gourieff at The Hague. The covering dispatch runs: "Au milieu des occupations nombreuses auxquelles les funestes événemens de Varsovie ont donné lieu, S.M.I. continue à vouer une attention suivie aux graves questions que le gouvernement des Pays Bas a consenti à soumettre

[11] Van Heeckeren to Verstolk, 13 Dec., R.A.; Bourgoing to Sebastiani, 11 Dec., A.é. Russie.

[12] Nesselrode to Lieven, 16 Dec., Ged., III, 434-7. Italics mine.

à la conférence de Londres." The Russian Minister is instructed to assure the Netherlands Government "que les événemens de Varsovie ne changent point l'attitude du cabinet impérial relativement aux affaires de l'Europe." [13]

These declarations could not, however, conceal to foreign observers the significant change which was reflected in the second point of the instructions to Lieven. The Netherlands Minister learnt of a similar dispatch on Belgian affairs, sent to Pozzo di Borgo, and attributed "la facilité avec laquelle la Russie est revenue sur sa première détermination de s'opposer constamment à l'indépendance complète de la Belgique avec une couronne séparée" – a misinterpretation, however, of the instructions actually sent to Lieven – to the difficulties caused to the Imperial Government by the Polish rising. He could derive but little comfort from Heytesbury's information that the dispatch to Paris with regard to Poland was concluded with a kind of threat, the less so as there was no such threat in the expressions employed by Nesselrode to Bourgoing in his verbal communication on the subject, nor in the account of it by Bourgoing, in whose accuracy the British Ambassador put little faith.[14]

Besides, Heytesbury too had been approached by the Vice-Chancellor with a request to write to London to prevent any pronouncement in Parliament "qui pût exciter l'esprit de révolte des Polonais et surtout rien qui pût attaquer le caractère personnel de S.M. l'Empereur." This was considered by Nesselrode as a great service. The Ambassador himself, in rendering the account of the Russian *démarche* to France for the second time, put his disapproval in still stronger terms. "This, it must be allowed, is a rather singular commentary upon the Emperor's ukaze from the beginning of November [announcing the Russian mobilization]," he wrote. "It betrays too much alarm by its complete contradiction to former language and shows too clearly the vulnerability of this Empire. It might have been impossible to conceal this secret very long but surely it was not prudent to proclaim it." But there were other elements to be considered in the new Russian policy, namely in Belgium, which did not need to be regretted. Heytesbury's descrip-

[13] Nesselrode to Gourieff, 16 Dec., Ged., III, 440.

[14] Van Heeckeren to Verstolk, 20 Dec., R.A.; Heytesbury to Palmerston, 19 Dec., F.O. Russia; for last instructions to Lieven prior to the Polish revolution see above, p. 55.

tion is almost a translation of Van Heeckeren's, quoted above. "After all declarations previously made", commented the Ambassador, "[it] speaks a sufficiently clear language", but, "it will greatly tend to facilitate an arrangement of Belgian affairs and diminish the chances of war." [15]

The British Ambassador enlarged on the situation inside the Russian Empire. No force capable of attacking the Polish army was to be expected at the Vistula before February. Hence, it was feared at St. Petersburg, the French might, in an attempt at a diversion in favour of the Poles, attack either in Belgium or on the Rhine. "The full occupation given to the Russian army by the Polish insurrection may indeed render the temptation of recovering former frontiers irresistible", thought Heytesbury. And in any case, the offensive power of Russia should now be considered as "more imaginary than real." Lastly, the Emperor Nicholas himself believed that the peaceable declarations of the French Minister would cease, giving way to a different attitude, as soon as he heard of the Polish revolution.[16]

As it happened, news of a still earlier date of a seemingly different line in French foreign policy were about to arrive in the Russian capital; it will be useful, however, first to recount two other minor incidents, which in another way illustrate the part played by the distance between the Western and Russian capitals.

During a Court reception the Emperor of Russia was said to have found a gracious word for every member of the diplomatic corps except for the French *Chargé d'affaires*, who was told: "Je n'ignore pas que ce qui vient de se passer en Pologne a eu sa source en France." This account passed through Berlin and in the latter half of December spread in the West; Grey, and even Mme Lieven believed and regretted this incident. In fact, much as he was disturbed by the Polish revolution, Nicholas liked Bourgoing, and the words he addressed to him were: "quant à vous, mon cher ami, je ne vous en parle pas, je sais... combien vous en soyez affligé; mais voilà ce que c'est quand une fois on commence." [17]

[15] Heytesbury to Palmerston, 16, 19 Dec., F.O. Russia.

[16] Heytesbury to Palmerston, 20 Dec., F.O. Russia.

[17] Chad to Palmerston, 17 Dec., F.O. Prussia; Talleyrand to Sebastiani, 28 Dec., Pallain, pp. 151-2; Grey to Mme Lieven, 29 Dec., Mme Lieven to Grey, 30 Dec., *The Lieven-Grey Correspondence*, II, 130-1; Bourgoing to Sebastiani, 8 Dec. 1830, 29 Jan. 1831, A.é. Russie; Van Heeckeren's account (9 Dec.) is also inaccurate.

A somewhat different type of evidence, and perhaps of a more serious character, is offered by the story of Sebastiani's dispatch of 22nd November – at the time of Langsdorff's mission to Brussels in favour of the House of Orange, and upon the receipt of the news of the Russian mobilization – concerning, among other things, the French support for the House of Orange in Belgium; this dispatch was destined for the eyes of Nicholas. At Nesselrode's request Bourgoing omitted the phrase saying that France "ne renoncera qu'en désespoir de cause" her support for William I and his family, from the copy to be sent to the Imperial Cabinet. On hearing of the omission – but not of the request – Van Heeckeren hastened to inform the Vice-Chancellor, who replied that he was satisfied with such documents as might be supplied by the French mission. The British Ambassador, however, learnt from Bourgoing that he had omitted another phrase also, referring to France's intention "of appealing in the event of war '*aux affections populaires*'"; in doing this he had been following private instructions from Sebastiani, "*to translate into the language of the Court where he was residing the plainer language of the constitutional Cabinet of Paris*". The Ambassador wondered if the French Minister would concur with such an interpretation of his confidential instructions, and concluded that the French dispatch, mutilated to this extent, could indeed have created the impression that the Russian mobilization had produced a desirable effect in Paris.[18]

Here again the British Ambassador seemed to forget that the interview of Bourgoing on the subject had taken place before the first news from Warsaw; Nesselrode may have been convinced that the omission of the phrase which would unduly anger his master, could do no harm since in view of the pressure of thePowers France would not dare to oppose the Orange cause in Brussels. When the question came to the knowledge of Heytesbury Nicholas' attention had been drawn to another matter: the intercepted Polish letter caused him to send for General St. Aldegonde, a Frenchman decidedly of the ancien régime, since a long time in Russian service. St. Aldegonde would, in the Emperor's opinion, be the man to answer the question: could the French Government withstand the clamour of the public and abstain from interference in Polish affairs? The reply was most

[18] Bourgoing to Sebastiani, 7 Dec., A.é. Russie; Van Heeckeren to Verstolk, 8 Dec., R.A.; Heytesbury to Palmerston, 8, 19 Dec., F.O. Russia. Italics by Heytesbury. For Nesselrode's request see above p. 55.

satisfactory: the French Government would loyally abstain from encouraging the Poles.[19]

Hardly two days had passed when Sebastiani's dispatch of 1st December – quoted in Chapter I – arrived in St. Petersburg. Even Bourgoing realized that to say in St. Petersburg that the Russian army must not even enter Prussia would be greatly resented by the Imperial Cabinet. He therefore began by referring to the question of the Belgian crown. The Palais Royal would like to place it upon the head of the Prince of Orange's son, if, as it seemed in Paris, his father could not count on success. But Nesselrode disliked the idea, as it involved a regency for a prince who was a minor. The King himself or the Prince of Orange was the only solution. The introduction not having been too successful, Bourgoing yet proceeded with his momentous declaration. As was only too likely, the Vice-Chancellor frankly showed that he felt offended. The statement "ne cadre pas avec les assurances amicales qui nous sont d'ailleurs parvenues", he said. "L'Empereur n'admettra pas une pareille condition et ne voudra jamais placer sur la même ligne un secours qui pourra lui être demandé par ses voisins et ses alliés, et votre sortie de chez vous. *Cinq cent Polognes* se fussent-elles révoltées, nous n'admettrons pas cette similitude." [20]

And though Nesselrode did not say so, and perhaps failed immediately to realize it, further reflections on the subject and especially consideration of the original date of Sebastiani's dispatch – 1st December – could not but unfold evidence of the "connaissance *préalable* qu'aurait eue le Ministère français des troubles survenues en Pologne; connaissance qui l'aurait déterminé à se dépêcher de faire cette notification pour ne pas avoir l'air de profiter des embarras où l'on croit la Russie actuellement engagée." It seemed, too, that Sebastiani had in the meantime realized that his declaration might produce an entirely undesirable effect at St. Petersburg and had requested the Prussian Government, always on a far better footing with Russia, to help remove this undoubtedly poor impression.[21]

Before replying to the French declaration Nesselrode invited the

[19] St. Aldegonde to Louis Philippe, 23 Dec., describing the audience on the 18th. A.é. Russie.

[20] Bourgoing to Sebastiani, 21 Dec., A.é. Russie.

[21] Van Heeckeren to Verstolk, 28 Dec., R.A.; Heytesbury to Palmerston, 25 Dec., F.O. Russia. Italics mine.

British Ambassador to an interview on the subject. "After a few remarks upon the extraordinary abruptness of such a declaration" the Vice-Chancellor developed his plan for a reply, in which reference was to be made to several instances of intervention in recent years, and he requested Heytesbury to write to England and try to counterbalance French pretensions. On 23rd December Bourgoing learnt of the Russian reply, in which the intervention in Belgium was not mentioned, but, observed the French *Chargé d'affaires,* this idea, "bien qu'elle soit toujours reléguée dans les cas éloignés et improbables", was still maintained by the Imperial Government. The Netherlands Minister called the Russian note "digne et modérée", but regretted the absence of any reference to Belgium, while Heytesbury noted a "more guarded language" than he had expected and none of the instances of intervention held out to him by the Vice-Chancellor. It will be noted that in the meantime the French *Chargé d'affaires* announced to the Imperial Government the special peace mission of the Duc de Mortemart, which Nicholas had called "une attention personnelle", and of which he would be very appreciative, while during those same days Polish envoys were being received in St. Petersburg.[22]

II

At the same time the influence of events in Warsaw made its way to Paris and London, where the arrangements to be introduced in the Netherlands were the problem of the day. While the King of the French saw the Polish rising as "une sauvegarde inespérée pour la France", and the diplomatic corps believed in the sincerity of renewed French pacific professions, there were dissenting voices; the Dutch Minister doubted if peace would be maintained.[23]

At the London Conference Talleyrand proposed to take advantage of the circumstances immediately. After achieving the suspension of hostilities, the Conference proceeded to try and bring about the re-opening of the Scheldt. The last protocol, the French Ambassador wrote home, would put the King of the Netherlands

[22] Heytesbury to Palmerston, 21, 24, 25 Dec., F.O. Russia; Bourgoing to Sebastiani, 23 Dec., St. Aldegonde to Louis Philippe, 23 Dec., A.é. Russie; Van Heeckeren to Verstolk, 28 Dec., R.A. cf. R. F. Leslie, *Polish Politics and the Revolution of 1830,* 1956, p. 143 sq.

[23] Apponyi to Metternich, 13 Dec., de Guichen, p. 243; Fagel to Verstolk, 13 Dec., R.A.; both are referred to again below, p. 86.

"au pied du mur", the more so as "les mouvements de Pologne lui ôtent une de ses espérances." An answer from The Hague was expected daily, fortunately "le tems presse: tout se fera mieux et plus facilement pendant les troubles de Pologne", he commented.[24]

The King of the Netherlands also received the friendly advice to raise the blockade. Bernstorff wrote about the difficulties caused to the Russians by the Polish rising, as well as about the troops which the Prussian Government felt compelled to send to watch the Eastern frontiers. The Prussian Minister advised the King not to reckon on the assistance of any single Power but to set all his hopes on the Conference alone. The Prussian Government tried to impress his conciliatory attitude on his British partner as well: Ancillon would not admit to Chad that Prussia might join Russia in suppressing the Polish rising and this corresponded with Chad's own belief that Prussia, in an anxious desire to avoid war, would remain neutral in the Polish as well as in the Belgian question.[25]

The leaders of British foreign policy also believed that negotiations with William I would be facilitated by the Warsaw events, and that in accordance with the little faith they placed in the Prussian declaration. In answer to the British Ambassador's letter describing an audience with the King, Palmerston remarked: "The language of the King... was certainly not equivocal, and proves that he must rely on Prussian aid; for upon his 6.000 Dutchmen alone he could certainly not have found so warlike a disposition. But if the accounts we hear to-day from Poland prove true, Russia and Prussia would have enough to do at home, without embroiling themselves with France for the sake of Holland." Sir Charles Bagot held the same opinion. A few days later, when the new Dutch plenipotentiary was leaving for London, it appeared impossible to the Ambassador for the Polish news not to influence Baron van Zuylen's attitude in London.[26]

However, this impression soon proved wrong. The Prussian Minister in The Hague obtained an audience of the King and found him in much better spirits, as if he had arrived at a decision. The King complained, though, that the Conference seemed now to have

[24] Talleyrand to Mme Adelaïde, 13, 15 Dec., *Nouvelle Revue Rétrospective*, (1901) XV, 232, 235.

[25] Bernstorff to Waldburg- Truchsess, 12 Dec., W. Gronemann, *Die Haltung Preussens in der belgischen Frage*, 1928, p. 14; Chad to Palmerston, 16 Dec., F.O. Prussia.

[26] Palmerston to Bagot, 10 Dec., Bagot to Palmerston, 14 Dec., Bagot MSS.

placed him on the same footing as his revolted subjects. He resolved, therefore, to save his honour by his own means, and that, the Prussian Minister thought, meant letting the Belgians have a touch of the winter with industry stranded. This would, of course, include keeping the port of Antwerp idle; hence the following impression of the British Ambassador: "The events in Poland have not had the effect that I anticipated – on the contrary, the argument now is that, in proportion as they deprive the King of the hope of foreign assistance, they throw him more upon his own resources, and make it more impolitic in him to surrender any advantage which he may now possess." [27] The Scheldt was consequently kept closed until January.

This, however, would not affect the London Conference. The suspension of hostilities already implied the recognition of the Belgians as a party, and Talleyrand pushed matters further, seeking to establish the principle of an independent Belgium. Less opposition was expected. "Il est évident que les événements du Nord portent leur fruit," noted the hostess of the French Embassy, "et que les esprits, si j'en excepte celui du roi de Hollande et de M. de Lieven, sont fort adhérents." Bülow and Wessenberg did not think any more of supporting the cause of Nassau, she thought, and were as anxious as the French to settle the Belgian business. Indeed, Bülow even feared that the Russian Ambassador's obstinacy in insisting on the rights of William I might bring about a *rapprochement* between England and France; Wessenberg's instruction allowed him great discretion but he spared no words to describe his reluctance in giving his consent to the protocol which recognized Belgian independence. But it was Lieven who presented the real difficulty, and even Talleyrand expected that he would not sign the Protocol but take the matter ad referendum.[28]

Palmerston, however, seized the proposition made to him by the French Ambassador as a means against a possible French *agrandissement* and took upon himself not only the preliminary talks with all the plenipotentiaries but also the formal authorship of Protocol No. 7, which was signed by all members of the Conference. The Foreign

[27] Allegri to Metternich, 16 Dec., Ged., III, 185; Bagot to Palmerston, 17 Dec., Bagot MSS.

[28] Duchesse de Dino to Mme Adelaïde, 18 Dec., *Nouvelle Revue Rétrospective*, (1901) XV, 236; Bülow to Frederick William III, 18 Dec., Gronemann, p. 37; A. Arneth, *Johann Freiherr von Wessenberg*, 1898, II, 97.

Secretary had to find a middle way between Talleyrand's insistence on employing the term independence, and the opposition of the Eastern representatives; Wessenberg, for once, proposed that more information should first be obtained from the Belgians. Out of these difficulties the expression "indépendance future" was coined; this greatly pleased Talleyrand and seemed the least offensive to the Eastern Powers. "It is a great step gained", wrote Palmerston to Grey, "to have got Russia and Austria to admit the necessity of early independence." [29]

At that moment, however, it was only Lieven, who had given way. "La signature de l'Ambassadeur de Russie m'était bien précieuse à avoir", Talleyrand prided himself, "et vous l'y verrez." As mentioned before, it had not been due to his effort: "The Conference went off extremely well", noted Palmerston. It seems that the expressed desire to keep the unity of the five Powers must have been the final argument which caused the Russian Ambassador to sign Protocol No. 7. New instructions which reached him well after 20th December allowed him to consent to the separation of Holland and Belgium only after William I agreed to it - a most unlikely result to be obtained not only at the time, but, as will be seen, even several months later – and thus virtually amounted to the exclusion of such an arrangement. On the other hand, though the Protocol of 20th December took account of the failure of the United Netherlands, it yet decided nothing about the sovereignty of the future Belgian State.[30]

In his report the Russian Ambassador defended himself by referring to French annexationist ambitions. Only a close unity of action with Britain – which he in any case had been instructed to maintain – could help in trying to avoid Belgium falling a prey to France. Lieven even added that the British Government itself had become so utterly dependent on the current of public opinion, which clamoured for a liberal policy towards Belgium, that it not only was far from being able to maintain the Treaties, but perhaps could even easily be made to use its resources, from which Russia had so much

[29] Palmerston to Grey, 19 Dec., Howick MSS.

[30] Talleyrand to Mme Adelaïde, 21 Dec., Talleyrand, *Memoirs*, III, 425; Palmerston to Grey, 20 Dec., Howick MSS; Van Heeckeren demanded in The Hague the granting of large powers to the Netherlands Ambassador in London in connection with the return of Lieven "dont le caractère timide et pusillanime est connu"; he considered it a very favourable circumstance that Matuszewic was to stay in London during the winter (to Verstolk, 17 Nov. 1830, R.A.).

benefited during the Napoleonic war, against all the Powers who were endeavouring to uphold these Treaties. The Russian historian who records this opinion feels unable to do so without attributing a prophetical character to it, bearing upon the Crimean war.[31]

The Protocol of 20th December came as a sudden blow to the Netherlands Government. In his reply to dispatches which had announced this unexpected decision of the Conference to him the Minister for Foreign Affairs, Verstolk, laid particular stress upon the unexpected consent of the Russian representative which, in his opinion, was in striking contrast to the views expressed by the Emperor Nicholas concerning the Polish rising. Speaking to the Russian Minister in The Hague Baron Verstolk said that the King's reply to the Protocol would be delayed until he had heard that the Courts of St. Petersburg, Berlin and Vienna approved the action of their representatives. In the meantime, the Dutch delegates formally protested against a decision of the Conference having been taken in their absence. The only comforting view which reached the Netherlands Government was that of Ancillon; in his dispatch, destined for William I, he said that the allies of the King only availed themselves of the expression *indépendance future* of Belgium to avoid the unconditional independence proposed by Talleyrand. Ancillon maintained that the decisions of the Conference had little binding power.[32]

The Netherlands Government felt unable to accept easily the tenor of Protocol No. 7, and the second plenipotentiary to the Conference, Baron van Zuylen van Nyevelt, who had only just arrived in London to join Ambassador Falck, began by paying a series of confidential calls on the representatives of the Eastern Powers, endeavouring to obtain a positive explanation of their abandonment of the rights of King William I. His first visits to Bülow and Wessenberg proved fruitless. More satisfactory for Van Zuylen was his conversation with the Russian Ambassador. In view of the warm feeling of sympathy with which Lieven spoke of the affairs of the Netherlands the Dutch plenipotentiary "realized how imperative were the circumstances which had made Lieven accept responsibility for the 7th Protocol." The Ambassasor further promised to try to carry the Conference with him in maintaining the sovereignty of the King of

[31] Lieven to Nesselrode, 24 Dec., Martens, *Recueil*, XI, 445.

[32] Verstolk to Falck and Van Zuylen, 28 Dec., Ged., IV, 387; Gourieff to Nesselrode, 28 Dec., Ged., III, 448; Ancillon to Waldburg-Truchsess, 5 Jan. 1831, Ged., III, 37.

the Netherlands in Belgium, but he did not conceal his uneasiness about British policy on this point. Van Zuylen, however, who had met Lieven during the latter's passage through The Hague late in November, found his opinions changed and asked the Russian Ambassador what consideration would be given to the interests of the people of the Southern Netherlands (which he, of course, considered identical with those of the King). The answer was little more than "we live from day to day" noted the Dutch envoy, and he wondered if that meant that Lieven and other representatives of the Eastern Powers had been promised the neutrality of Britain and France in Polish affairs in return for their passive attitude as regarded Belgium. There is little evidence to determine whether he had any reason to make such a conjecture. In the course of his letter he further says that Esterhazy considered Austria also compelled to accede to Protocol No. 7 by the Polish events, and he attributed to the Austrian Ambassador the opinion that Lieven had been assured of English and French neutrality with regard to the contest on the Vistula and in this way brought to sign.[33]

If these findings indicated the source of the trouble, a more striking example of the influence of Warsaw events on the Belgian policy of the Powers Van Zuylen found in his second interview with the Prussian Minister. "Polish affairs have considerably transformed the respective standpoints of the representatives [at the Conference]", runs the *précis* of Van Zuylen's conversation, "the three Powers thought themselves more engaged, unlike France, who found herself stronger than ever, through her real importance, but, above all, by the force of her moral influence." To support his arguments Bülow communicated to the Dutch plenipotentiary the first draft of what was later to become Protocol No. 7 made by Talleyrand. This project was indeed based on entirely different propositions. It not only implied the loss of sovereignty in Belgium by the House of Orange, but declared also that the Belgians themselves had achieved their independence. Bülow thought that this draft was only a weak reflection of the feeling in France, and that it was preferable to declare Belgium independent than to expose her to the inevitable annexationist ambitions of France.[34]

Less apparent, perhaps, was the change which the direction of

[33] Van Zuylen to Verstolk, 24 Dec., R.A.

[34] Van Zuylen to Verstolk, 28 Dec., R.A. For the text of Protocol No. 7 and of that of Talleyrand's project see Appendix III.

Belgian affairs underwent in Paris after the Polish news. It first of all involved Luxemburg, which presented France with a two-fold problem. Firstly, the Grand Duchy was a family possession of the King of the Netherlands which he had obtained in 1815 in exchange for his Nassau lands; it belonged to the German Confederation. But the majority of its inhabitants sided with revolutionary Brussels, and while there was no dispute about the rights of William I with regard to Luxemburg inside or outside the Conference, it was equally obvious that an occupation, or rather, in this case, a conquest of the Grand Duchy would hardly be allowed by France, and all Conference Powers were aware of this. Secondly, some places lost by France in 1815 – of which Bouillon was one, administratively included in the Grand Duchy – were now claimed by French public opinion. It soon became obvious that the Government was not averse to the idea.

It had been understood for some weeks that the Federal troops would occupy the Grand Duchy on orders given by the Federal Diet. The vote of excluding Nassau, pronounced by the Belgian Congress on 24th November, speeded the arrangements, and the Conference acknowledged them in its protocols of 9th and 14th December. These protocols arrived in Paris after the news from Poland, and Sebastiani then wrote to Frankfurt to ask for a delay in effecting the occupation, under the plea that nothing had been decided about the fate of Bouillon, for which a special arrangement was necessary. Not much pressure was required, however, because after the Polish news Prussia, always reluctant to assent to such a dangerous measure, was less than ever likely to push the matter through the Federal authorities.[35]

A less important, though much more obvious *volte-face* was effected in the French tone with regard to the King of the Netherlands. On 7th December Sebastiani still assured the Netherlands Minister in Paris of "des bonnes intentions du Roi des Français et de son gouvernement dans la question de Belgique." Entirely different terms were employed by the French Minister three weeks later, after the Protocol of 20th December. Speaking of Van Zuylen's recent arrival in London he regretted that instead of bringing the news of the King's concurrence with the measures adopted by the Conference, he had reported that the Scheldt would be kept closed, thus putting

[35] W. von Franqué, *Luxemburg, die belgische Revolution und die Mächte,* 1933, pp. 39 and 44.

new obstacles in the way of a settlement. Sebastiani permitted himself to add that this policy gave evidence of the King's desire for war, which would not be favourable to his cause "depuis que les Puissances l'avoient en quelque sorte *forcément* abandonné...", and even went so far as to warn: "s'il persistoit dans son système, on seroit obligé d'en venir à des mesures plus positives que celles, dont on est convenu jusqu'ici." [36]

Only a few days later in trying to solve the Luxemburg problem according to the as he thought new political configuration, Talleyrand went much further than Sebastiani did here, and in this way he immediately introduced considerable coolness in the Anglo-French relations.

III

Though the King of the Netherlands had not accepted the Protocol of 20th December, the Belgians, trusting in the support of the Western Powers, were already trying to find a Royal personage willing to accept the vacant throne, and able to bring about some stability in their country. There were voices in Belgium for the Duc de Nemours, the younger son of Louis Philippe, and for the Duc de Leuchtenberg, but the provisional government of Belgium realized that the Conference Powers were very unlikely to accept either of these candidates. But there was another candidate who could command popular support whilst he was also put forward by both the British and the French. This was Prince Leopold of Saxe-Coburg, widower of the English Crown Princess Charlotte, who had been resident in England for many years. Talleyrand mentioned him early in December, while Palmerston let it be known to Lord Ponsonby, representing the Conference at Brussels, that, unofficial as it was, "it must end in Leopold." [37]

But Russia, through Lieven and Pozzo, was continually declaring her opposition to Prince Leopold, being moreover committed to the sovereignty of the House of Orange in Belgium as was

[36] Fagel to Verstolk, 7, 27 Dec., R.A. Italics mine. Reference is made in the second dispatch to "les arrangemens que les Plénipotentiaires de cinq puissances, réunis à la Conférence de Londres, jugeoient nécessaires à la gravité des circonstances actuelles, que les événemens en Pologne avoient rendus presqu'irrévocables." It is not clear whether the last qualification came from Sebastiani or from the writer of the dispatch.

[37] Palmerston to Grey, 18 Dec., to Ponsonby, 24 Dec., Howick MSS.

well known. Unexpectedly, however, when the Belgian delegation came to Paris late in December to seek support for Leopold's candidature from the French Government, Sebastiani refused to commit France in this question. Time was working for Belgium in realizing her wishes, he told the delegates. The Nemours crisis some weeks later seems to have brought evidence of the real meaning of this pronouncement.[38]

This was, however, much more than the King of the French as well as his London Ambassador considered feasible. "Nous serions parvenus, et avec beaucoup de peine", he wrote to the Minister for Foreign Affairs on 3rd January, "à obtenir la souveraineté de la Belgique pour le Prince Léopold, marié à une Princesse de France"; but that would require France to show good will, and first of all, formally to refuse the candidature of the Duc de Nemours, since otherwise "...la Russie, toujours prête à s'emparer de la politique de l'Angleterre, profiterait de cette circonstance [Nemours' candidature] pour pousser à l'extrême les hostilités de société qui ont ici une grande influence."[39] Unfortunately for France, there were other difficulties as well.

At this time Louis Philippe considered not only Russian opposition but the state of Belgium itself as factors responsible for the small chance Leopold might have; the King was prepared to accept any solution which would put an end to the Belgian troubles. "Il ne m'est pas démontré que ce ne sera pas pour le Prince d'Orange", he wrote to Sebastiani.[40]

There seemed to be no one, however, to press the Prince's cause inside or outside the Conference. The Russian representatives certainly, apart from Princess Lieven as will be shown, did not share their monarch's devotion to the House of Orange. Gourieff from The Hague considered its cause hopeless. He pointed out the "petites vues" of William I, the loss of popularity suffered by the Prince of Orange, finally – as if it could have been otherwise – the insincerity of France in her professed support of the Prince. As Sebastiani did to the Belgians, he proposed to his government to let time solve

[38] Lieven to Nesselrode, 7 Jan., Martens, *Recueil,* XI, 550; Gendebien to the Belgian Government, 29 Dec., C. Buffin, *Mémoires et documents inédits sur la révolution belge et la campagne de dix jours,* 1912, II, 545. This point is stressed by H. T. Colenbrander, *De afscheiding van België,* 1936, p. 78.

[39] Talleyrand to Sebastiani, 3 Jan., *Revue des Deux Mondes,* (1910) LV, 281, to Mme Adelaïde, 5 Jan., *Nouvelle Revue Rétrospective,* (1901) XV, 338.

[40] Louis Philippe to Sebastiani, 2 Jan., *Revue des Deux Mondes,* (1910) LV, 278.

Dutch difficulties. He would rather see a speedy crushing of the Polish rebellion – though, and perhaps because, he did not underrate the hardships of this struggle – where real Russian interests were involved.[41] Lieven, as mentioned above, was no match for Talleyrand, and Matuszewic, the other supposed defender of King William I's interests and by far the ablest of the Russian members of the Conference, did not show much zeal. His position as a Pole rendered it difficult for him to use the argument of the Polish revolution. To Van Zuylen he stressed the menacing attitude of France and Bülow's readiness to follow Palmerston, factors which, backed by his instructions to keep the five Powers together, constituted the reasons for his share in allowing the Protocol of 20th December.[42]

The initiative of an Orange restoration was taken, with considerable reluctance, by Palmerston. His and Grey's opinion on this subject, expressed in the dispatch to Lord Heytesbury at the end of December, was of course influenced by the opinions of the Belgian delegates and of Lord Ponsonby: to try for an Orange restoration would mean to work for a civil war. They were also influenced by their view of the personality of the Prince of Orange; but they no doubt realized equally well that only through showing that a civil war in Belgium must necessarily mean a general war could they achieve some success in trying to make St. Petersburg renounce its legitimist views with regard to Belgium.[43] Yet only a few days later Palmerston showed that he had his reasons in attempting to bring together the wishes of the Belgians and those of William I.

On 2nd January the Netherlands plenipotentiaries had an interview with the Foreign Secretary in order to acquaint him with the views of their King with regard to the separation of Holland and Belgium. Then Palmerston asked them, "in deepest confidence", whether the King was interested in the choice of a sovereign for Belgium, and whether he would consent to cede Luxemburg to the Prince of Orange, if his election proved possible. Palmerston added

[41] Gourieff to Nesselrode, 28 Dec., Ged., III, 449-50.

[42] L. Sapieha, *Wspomnienia* (Souvenirs), 1912, pp. 112-14. Another Polish counsellor of the London Embassy, Sobolewski, had left his post, much to the annoyance of Matuszewic. Mme Lieven herself made an effort to persuade Sobolewski to stay; Van Zuylen to Verstolk, 31 Dec., R.A.

[43] Palmerston to Heytesbury, 31 Dec., F.O. Russia, and priv., Heytesbury Papers, cf Webster, I, 122; Palmerston to Ponsonby, 24 Dec.: "I fear from all you say & I hear from other quarters that there is no chance for the Prince of Orange without civil war, & not much with it." Howick MSS.

that he was perfectly aware of the many difficulties connected with such an arrangement. The Dutch plenipotentiaries took this proposal, repeatedly described as his own by Palmerston, ad referendum. They asked the Foreign Secretary, however, if he did not think that France would oppose this combination as it represented a possible settlement of Belgian affairs; the Minister replied that on the contrary he feared that France would consider it a provisional measure, agreeable to her till a more convenient occasion would enable her to annex Belgium, and that she would support its acceptance on that ground. The Dutchmen, naturally enough, expressed their doubt as to the desirability of a solution offering perhaps such a prospect. "Palmerston admitted that there was a considerable party for the Prince, but it was weak and passive, whereas men of ability belonged to the opposite party, which by now had been strengthened by the deposition of the House of Orange that had already taken place. He further reviewed the almost impassable difficulty that the Prince would experience in administering Belgium and the necessity to bring up his son in the Roman Catholic religion. This idea of the Prince of Orange ruling in Belgium did not prevent him also from advising H. M. the King of the Netherlands, in the strongest terms, to apply every measure to restore order in Luxemburg, especially to announce an amnesty and to promise a speedy new constitutional arrangement." In their subsequent dispatch of the same date the envoys wrote: "The English Minister repeatedly drew our attention to the provisional character of this information, and to the doubts whether his ideas would find enough support with the others for them to be adopted. He did not wish to accept any responsibility for this whatever." This dispatch again stressed the importance Palmerston attributed to the measures he proposed with regard to Luxemburg.[44]

These confidential proceedings were not unknown at least to the most interested parties in the Conference. Princess Lieven knew that Palmerston was considering the plan of making King William I agree to cede Luxemburg to his son; yet she was not pleased with the attitude of the British Ministers towards the Prince of Orange, and she noted the existence of personal prejudices against him; nor was she herself blind to his faults.[45]

It should be noted to what extent the views of William I coincided

[44] Falck and Van Zuylen to Verstolk, 4 Jan., Ged., pp. L and 402.
[45] Mme Lieven to Alex. Benckendorff, 2 Jan., Lie MSS.

with those of the British Foreign Secretary, a fact which seems largely to have escaped the attention of historians. It was known that in his opposition to the Conference the King relied upon the support of Russia, though even before the outbreak of the revolution in Poland his principal advisers did not think any substantial military assistance from the East likely. On the other hand, the participation of Russian plenipotentiaries in proclaiming the independence of Belgium hurt feelings at The Hague very much; it seemed unlikely that this proceeding could obtain the full approval of St. Petersburg – which again was not the same as a disavowal of Lieven's and Matuszewic's signatures.[46] Lastly, William I felt unable to agree to the state of things in his Southern provinces after the revolution; yet he realized the difficulties in continuing, or rather, as the situation was, in re-establishing his direct authority there.

Minister Verstolk's instructions for the Dutch envoys in London, dated 31st December, 1830, were the product of the above considerations. The King desired them to propose to the Conference that Belgium, separated from the Northern part of the Kingdom, should remain under his sovereignty but be governed independently by the Prince of Orange, as Governor-General or Lord-Lieutenant. Verstolk rightly pointed out that however far the Protocol No. 7 might have gone in its disregard of the rights of the King, it did not prejudice the question of the sovereignty of Belgium. These instructions were supported by appendices, viz., a letter from Sebastiani to Bourgoing, professing the desire of France to see the Prince of Orange on the Belgian throne, and Nesselrode's dispatch to Gourieff, of 16th December, already referred to above. It is of course improbable that the Dutch Government was unaware of the motives which had prompted Sebastiani to make such a declatarion, motives which no longer existed; but it could safely assume that France had committed herself by this statement.[47]

Falck and Van Zuylen made those instructions the subject of their long report of 5th January, 1831. Lieven and Matuszewic, having expressed their entire concurrence with the project as one which would particularly please their Imperial master, observed at

[46] Verstolk and Van Maanen to William I, 17 Nov., 1830, Verstolk to Falck and Van Zuylen, 28 Dec., Ged., IV, pp. 303-4 and 387.

[47] Verstolk to Falck and Van Zuylen, 31 Dec., Verstolk Papers, R.A. A major part of this document is printed in Ged., IV, p. L.

the same time that, unfortunately, "le Cabinet Britannique regarde cette idée comme impossible à réaliser dans les circonstances actuelles. Y insister, ou seulement la proposer à la Conférence aurait pour résultat infaillible de rapprocher encore d'avantage l'Angleterre et la France." In consequence, in a general war Britain would be on the side of France.[48]

Curiously enough, at this very time Princess Lieven was still obtaining Palmerston's assurances that he was continuing his efforts in favour of the Prince of Orange – though he himself did not seem to place any faith in them; nor did Palmerston refer to any difficulties made by the French Government but only to the Belgian delegate Van de Weyer's opinion "que le Prince d'Orange c'est la guerre; que le Prince Léopold c'est la paix éternelle..." Nothing, however, had been done by the Russian delegation to reconcile William I's opinion with the possibilities offered by their influence and their knowledge of the situation.[49]

Instead, it was Palmerston again who took up the subject. On 2nd or 3rd January, about the same time that he had his conference with the Dutch plenipotentiaries, he spoke to Talleyrand about the Belgian problem and proposed to make the King of the Netherlands agree to cede Luxemburg to his son on condition that the latter was accepted by the Belgians – an arrangement, he thought, "which might or might not prove practicable but which if it could be accomplished might remove many difficulties." "Talleyrand looked very grave", observed Palmerston; instead, however, of referring to the difficulties of placing the Prince in Belgium, he referred to those which would be made by the French Government in connection with the strategic importance of the Grand Duchy. This could not be a greater obstacle in the case of Belgium than it was in the case of the United Netherlands, Palmerston observed. But Talleyrand wondered if an arrangement might prove possible by which Luxemburg were given to France. "I confess I felt considerable surprise," wrote Palmerston to Granville a few days later, "at a proposition so much at variance with all the language and professions which he and his Government have been holding." He then learnt that Talleyrand had sounded the Prussian Minister on the subject of exchanging the Rhenish provinces for Saxony, in

[48] Falck and Van Zuylen to Verstolk, 5 Jan., R.A.

[49] Mme Lieven to Alex. Benckendorff, 5 Jan., Lie MSS; cf. Palmerston to Ponsonby, 4 Jan., Palm MSS. See Appendix IV.

which case the King of Saxony would be "transferred" to Brussels; and another proposal was made to himself, that France would like to obtain Philippeville and Marienbourg in return for her support of Prince Leopold's candidature. The interview with the French Ambassador with regard to Luxemburg marks, therefore, a turning-point in Anglo-French relations. "This is the first time that he [Talleyrand] has let out any distinct proposition for adding to the French territory", wrote Palmerston, and he commented later: "It looks as if France was unchanged in her system of encroachment, and it diminished the confidence in her sincerity and good faith which her conduct up to this time had inspired." [50]

For the Russian mission in London Talleyrand's slip was a welcome one. Princess Lieven, who made such a point of speaking to Grey and Palmerston of the Prince of Orange's return to Brussels, that "ni l'un ni l'autre n'a osé encore me nommer le Prince Léopold que comme une idée très en l'air", fully perceived the importance of Talleyrand's having "singulièrement bien secondé" her efforts. She also claimed to have proved to Grey that the Belgian delegate Van de Weyer "a haussé le ton", trying to establish that the Belgians wanted to belong to France, or at least to have a French prince – an obvious reference to the Duc de Nemours – after he had conferred with the French Ambassador. "Lord Grey est furieux," noted the Princess, adding that Grey's insistence on a decisive movement on the part of Russia in Poland now appeared stimulated afresh.[51]

Even before this memorable and momentous interview the Belgian delegate in London believed that any movement in favour of the House of Orange would render Russia unwilling to agree to any other arrangement. Louis Philippe, as has been mentioned, no doubt *à contre-coeur*, seriously considered the possibility of an Orange

[50] Palmerston to Ponsonby, 4 Jan., Palm MSS, to Granville, 7 Jan., Bulwer, II, 27-9. There is not enough evidence as to whether this interview took place before or after the interview with the Dutch envoys. In his letter to Ponsonby Palmerston speaks of "the day before yesterday". Mme Lieven, in her notes on revolutionary events of 1830–31, placed this interview on the 3rd (Lie MSS 47377); Bülow's account, dated 5 Jan., adds nothing to clarify the question, cf. Franqué, p. 260.

[51] Mme Lieven to Alex. Benckendorff, 2, 6 Jan., :"Je vous ai déjà mandé ce qu'il [Grey] m'a dit du premier mouvement, il les répète aujourd'hui, les mêmes paroles: frappez, fort et vite, et il fait tous les voeux pour que l'Empereur se rende maître de cette révolution." Lie MSS. This passage is preceded by one describing the unfortunate attempt of Talleyrand.

restoration because of the internal situation in Belgium, and no other view was held by Van de Weyer: Belgium was ripe for such a solution. "Ne nous dissimulons point," he wrote to his colleague in Paris, "nous sommes sur un volcan; le pays est sur le point de nous échapper; les partis s'agitent; l'Orangisme fait des progrès; incertitude sur notre avenir y jette beaucoup de monde." [52]

As may have been expected, therefore, Lord Ponsonby, the British member of the Conference's delegation to Brussels, also began to observe there "a crisis which if taken advantage of, may ensure the restoration of the Prince of Orange." He soon came to the conclusion that if the Prince could bring Luxemburg with him, his election in Belgium would be certain.[53]

Under the influence of his interview with the French Ambassador Palmerston was of course only too glad that the idea he had conceived seemed realizable. He could reckon on the whole-hearted support of the representatives of the Eastern Powers. Talleyrand, after his discomfiture with regard to Luxemburg, thought it better not to oppose the rest; in letters home he concealed his weakened position by attributing some influence in this question in London even to the Prince of Orange himself.[54]

The Conference which at that time was mainly engaged in studying projects for the arrangements for the separation of Holland and Belgium, delegated Lieven, Matuszewic and Wessenberg to approach the Prince and ask him to issue a proclamation to the Belgians. The draft of this document had been written by Matuszewic but was subsequently altered by Grey to such an extent that, according to Talleyrand, "c'est lord Grey qui en est le véritable auteur. Il y met un tel prix et un intérêt tel que nous n'avons pas pu nous opposer à ce que cette tentative fût essayée." Yet Talleyrand himself also helped with the preparation of the document and reckoned with the possibility of its success.[55]

Grey thought of yet another measure. He realized that Ponsonby and Bresson, the British and the French member of the Conference's delegation, could not assist with the proclamation. But the proclamation by itself was not enough: "A communication

[52] Van de Weyer to Gendebien, written a day or two before 29 December 1830, Th. Juste, *Alexandre Gendebien*, 1874, pp. 45-6.
[53] Ponsonby to Palmerston, 6, 7 Jan., Palm MSS, 11 Jan., Webster, I, 122.
[54] Talleyrand to Mme Adelaïde, 12 Jan., Talleyrand, *Memoirs*, IV, 12.
[55] Talleyrand to Mme Adelaïde, 16 Jan., *Nouvelle Revue Rétrospective*, (1901), XV, 343.

made by him, *if not very carefully looked after,* may spoil everything", he wrote to Palmerston. Could there be no invitation from the Prince's influential friends? [56]

Palmerston had a plan of a slightly different nature. Through the intermediary of Bagot he requested the Papal Nuncio in the Netherlands, Mgr. Capaccini, who had spent most of the preceding three months in London and, after a brief stay in The Hague, was about to come to England again, to use his influence with the Belgian clergy in favour of the Prince. The Foreign Secretary had seen the Nuncio before and found him kindly disposed to the cause of the House of Nassau; he thought therefore that the Nuncio would be willing to comply with this request. But Mgr. Capaccini had in the meantime left The Hague and Palmerston's letter could not be delivered to him.[57]

The sending from London of the proclamation of the Prince which was to be distributed by the Prince's influential friends in Belgium, had already taken place in the greatest secrecy in order to avoid offence to the French *parti du mouvement* as well as the impression of the Conference interfering with Belgian internal affairs. The King of the Netherlands was only informed after the attempt had been agreed on, and the preliminary steps had been taken. Even the Prince of Orange reported later to his father: "Je n'y suis que pour ma signature." While this was going on, Lord Ponsonby came to the conclusion that "the chances of the Prince of Orange have increased more than in geometrical ratio." [58]

Van Zuylen – who for the rest knew nothing of these moves – again approached the Russian Ambassador. Stubbornly as he had defended the rights of the King of the Netherlands before 20th December, Lieven, once overridden, in the same way believed in the new situation, and had a flow of arguments to support him. As Belgium was the chief concern of England and France, he said, one could not insist on principles such as treaties or sovereign rights, because this would lead to conflict with France. "I looked in an entirely different way at the problem at the time of my passage through The Hague, and I am sorry that I then envisaged pros-

[56] Grey to Palmerston, 10 Jan., Howick MSS. Italics mine.

[57] Palmerston to Bagot, 12, 13 Jan., Bagot MSS, to Grey, 13 Jan., Howick MSS. See Appendices V and VI.

[58] Webster, I, 123; H. T. Colenbrander, *Willem II, Koning der Nederlanden,* 1938, II, 82-4; Ponsonby to Bagot, 11, 14 Jan., Bagot MSS.

pects which do not exist," he proceeded. "No good is to be expected from a war. Views from St. Petersburg cannot be considered here; after any matter has been decided upon here, 5 to 6 weeks must pass before an answer can arrive from St. Petersburg." His final declaration was still more positive. "It is not the first time that I must act against my instructions," the Ambassador said, "but I am able to answer for it, and I am certain of the approval of my sovereign, who, were he here, would have given the same advice to your King." It will be shown that this statement presented a striking contrast to that made a few days later by Nesselrode to the French *Chargé d'affaires;* nor did Lieven mention the circumstance which he knew had changed the attitude of the British Cabinet.[59]

He did do other things, however. The Russian Ambassador's activity in Belgium consisted in maintaining an agent, an old Russian diplomat, Baron Krüdener; Ponsonby contacted him as well. It was Lieven who informed Palmerston of the arrival of an Orangist delegation from Ghent. But Palmerston found the evidence of the Orangist movement, collected by Lieven, insufficient, and he also expected that William I would not allow his son to accept the crown of Belgium. Yet Palmerston was prepared to try to carry out the Prince's election: "...et le Roi s'y soumettra, de mauvaise grâce, comme à l'ordinaire." [60]

Palmerston accordingly thought he had found an opportunity to enlist support for the Prince by seeing the Papal Nuncio on 18th January, immediately after Mgr. Capaccini's arrival in London. The Foreign Secretary renewed his request to the Nuncio to intervene with the Belgian clergy on the Prince's behalf. But Mgr. Capaccini declined: the election of the Prince could only be made possible by a dissolution of the Congress, where the Catholics had a majority, and he, the Nuncio, could not advocate this measure which he considered would necessarily bring about another revolution. He moreover did not believe in the Prince's success, knowing that the Prince was not prepared to give up his rights in Holland. Yet Palmerston expressed the hope that the election might prove possible without the dissolution of the Congress.[61]

[59] Van Zuylen to Verstolk, 18 Jan., R. A. Cf. below, p. 95.
[60] Palmerston to Lieven, 18 Jan., Lie MSS.
[61] E. Perniola, "De internuntius Mgr. Francesco Capaccini en de Belgische omwenteling van 1830", in *Mededelingen van het Nederlandsch Historisch Instituut te Rome,* Third Series, IV, (1947), pp. 153-158.

At this time, however, unwelcome news was already on its way from The Hague to London. The King of the Netherlands, after some hesitation, refused his consent to the arrangement proposed by the Conference. The Conference was by then about to complete its other great Protocol, No. 9, of 20th January, which guaranteed the neutrality of Belgium; Belgium having thus been made safe from the immediate annexationist ambitions of France – Talleyrand's opposition to the Protocol is the best evidence of this – the Orange candidature was for Lord Grey's administration no longer of paramount importance. But there is not enough evidence that the British interest in an Orange restoration was purely momentary. The Dutch plenipotentiaries, both opposed to this measure, believed that all the British Ministers sincerely supported the Prince. Grey and Palmerston may have had their doubts as to the desirability of an Orange restoration in the person of the Prince, yet they appreciated that this would be a radical step in preventing new French intrigues in Belgium. It was up to the Russian representatives to try to reconcile the divergent views of the British leaders and William I; they made no attempt to do this.[62]

It will be seen in the following chapter that the Orange candidature was a nightmare – real or pretended – for the French Government up to the beginning of April. It remains to be noted here that as soon as the first sign of a favourable disposition on the part of the Conference in favour of the Prince was shown, Talleyrand warned his government of the existence of a strong party for the Prince in Flanders. Even before official steps had been taken in Paris via the British Embassy Talleyrand called on Grey to express the great disapproval of his Government with the Conference's interference in favour of the Prince. Whether the French protest was dictated by fear of offending the *parti du mouvement* – or – as Princess Lieven suggested – by the hope that the unsettled state of Belgian affairs would in the near future allow a French annexation of that country, Grey did not know; he did have, however, a positive opinion as to the situation which made the French protests possible. "If you, and Prussia, and Austria were not all at this moment *les mains liées*", he wrote in answer to Mme Lieven's letter, "we might soon settle this

[62] Bagot to Palmerston, 18 Jan., Bagot MSS. See Appendix VII; A. R. Falck, *Gedenkschriften* (R.G.P. Kleine Serie, Vol. 13), p. 312, cf. Falck to Van Lennep, 24 Jan., A. R. Falck, *Ambtsbrieven*, 1861, p. 308; Van Zuylen to Verstolk, 21 Jan., Verstolk Papers, R.A.; for Palmerston's opinion see Webster, I, 121-2.

business." But there was as yet no such hope, he observed, since dispatches from St. Petersburg announced that no adequate forces could be expected on the Vistula before the end of February.[63]

IV

The assumption made by Van Zuylen on the Russian motives for their share in the Protocol of 20th December seems to have hardly any foundation. There is no evidence that either Talleyrand or Palmerston made any reference to Polish affairs in the conversations preceding the signing of the Protocol. There is, however, a possibility that either of them, especially Talleyrand, may have done so and – even in very vague terms – availed himself of the well known weakness of Lieven. It was the French Ambassador who on 21st December devoted a fairly long dispatch to the Polish commotions. He surveyed the course of Polish history as witnessed by himself and expressed his regret that Napoleon had failed to re-establish the Polish state. "Il n'est personne maintenant", he wrote, "qui ne comprenne que le royaume de Pologne, fortement constitué, formerait la meilleure barrière contre les envahissements menaçants de la Russie." He reckoned that there would be sufficient national forces inside the former Polish territories, as well as supporting factors inside and outside the Russian Empire, to permit the creation of a Polish State, "si l'Angleterre voulait entrer franchement dans nos vues." Moreover, he believed that this could be achieved without war: "le cabinet de St. Pétersbourg, bien conseillé, céderait peut-être avec le temps à des démarches habilement combinées." Perhaps it was one of these, as yet entirely unofficial, *démarches* which contributed to make Lieven give way in the matter of Protocol No. 7, a circumstance, according to Van Zuylen, not unknown to the Austrian Ambassador.[64] It will be seen in the course of this section as well as in the following chapters that Talleyrand himself successfully used the Polish argument, though not entirely for the benefit of the Poles.

As soon as the first news of the Warsaw riots reached Paris,

[63] Talleyrand to Sebastiani, 12 Jan., A.é. Angleterre; Grey to Mme Lieven, 17, 18 Jan., Mme Lieven to Grey, 18 Jan., *The Lieven-Grey Correspondence,* II, 136-9.

[64] Talleyrand to Sebastiani, 12 Dec., Pallain, p. 143; cf. Van Zuylen to Verstolk, 24 Dec., referred to above, p. 72.

Prince Leon Sapieha, a Treasury official of the Kingdom of Poland, who at that time was staying in the French capital, went to see Sebastiani to learn the Minister's view on the subject. As there was no definite opinion as to the character of the movement, the French Minister said that if it were mere street rioting the best way out of it was a settlement by St. Petersburg alone. But if it appeared to be a national movement, the Poles might rely upon the support of France. Sapieha then asked if he might communicate this statement to Prince Adam Czartoryski (his brother-in-law).[65] To this Sebastiani agreed. He further advised the Pole to go to England and to try to enlist the support of Lord Grey's administration.[66]

At the same time the French Minister wrote a dispatch to St. Petersburg. Uncertain of the extent of the Polish movement, he expressed his concern and also his apprehension that, even if it spread beyond Warsaw, the Russian forces might still be able to suppress it. For the rest, the policy of France would in no case deviate from its peaceful direction; the Government was anxiously awaiting news from St. Petersburg and especially the arrival of credentials for Pozzo di Borgo as these were already more than two months overdue, so as to be able to confirm its intention of sending the Duc de Mortemart (who before the July revolution had represented France in the Russian capital) on his peace mission to the Emperor.[67]

The French Cabinet was wise enough to renew their pacific assurances and the Austrian Ambassador reported the unanimous agreement of the diplomatic corps on their "loyauté et courage." The Dutch representative, however, was far from believing that the warlike feeling in France had diminished. Besides the usual arguments that Russian forces were being employed in Poland, and therefore unable to help to check the spreading of the revolution in the West of Europe, Baron Fagel observed that the war party, stimulated by this circumstance, had "un appui décidé dans un parti du Ministère français." [68]

[65] Prince Adam Czartoryski was the former friend of Emperor Alexander I and for a time Russian Minister for Foreign Affairs. He was the most authoritative personality in Poland before 1830, and the obvious choice as head of the Polish Government, which he, with considerable reluctance, became after the *déchéance* of Nicholas, pronounced by the Polish Diet on 25th January, 1831. His recent biography is by M. Kukiel, *Czartoryski and the European Unity*, 1955.

[66] Sapieha, *Memoirs*, p. 111.

[67] Sebastiani to Bourgoing, 11 Dec., A.é. Russie.

[68] Cf. note 23, above, p. 67.

There were other factors as well which might warrant doubts with regard to the intention of the French Government. The French re-armament continued and was the subject of diplomatic activity during December, 1830; comments in Vienna and Berlin have already been mentioned above. Immediately after the Warsaw news the British Ambassador in Paris, Lord Stuart de Rothesay, proposed to the French Government a suspension of its military preparations, trying to persuade it that no dangerous measures from Russia would now arise. It seemed to the Ambassador that Sebastiani was prepared to accept this view; but Louis Philippe believed that the continuation of re-armament "will produce the best security for the maintenance of peace." On 24th December Sebastiani repeated to Lord Stuart his fear that a bad impression would be produced on French public opinion by the suppression of the Polish revolution which, as he had already told the British Ambassador, was to be expected. The Minister instructed Talleyrand to enquire of the British Government if it would be inclined to offer to mediate in the contest between Nicholas and his Polish subjects before actual hostilities took place, Sebastiani talked about this *démarche* to the British Ambassador and stressed again his conviction that the Poles had no chance of success.[69]

The interest which other Powers continued to take in French re-armament made Sebastiani inform the Prussian *Chargé d'affaires* in Paris, Baron Werther, that the French Government planned to reduce its armaments until a favourable reception would be accorded to the peaceful declarations which the Duc de Mortemart was instructed to make in St. Petersburg; the ultimate news from the Russian capital would, he hoped, make it possible to suspend them altogether. The British Ambassador learnt of this interview from the Prussian representative but himself found that Sebastiani "held very different language to me, saying that it is useless to conceal from me his conviction that it is not expedient to suspend these preparations for war until he shall ascertain the course which the suppression of the Polish insurrection will induce the Russian Government to pursue." [70] It should be noted that only a few days later, on 8th January, Pozzo di Borgo presented his new credentials to the King of the French.

[69] Stuart to Palmerston, 10, 13, 17, 24, 27 Dec., F.O. France; Sebastiani to Talleyrand, 25 Dec., Pallain, p. 151.

[70] Stuart to Palmerston, 31 Dec., F.O. France.

It was about this time that the Poles – Sapieha, Count Walewski, and Marquis Wielopolski – came to sound the attitude of the British Government towards events on the Vistula. The London press was most sympathetic about the Polish movement, but less positive evidence is available on the views held in Society. Greville took a rather serious view. "The affair at Warsaw seems to have begun with a conspiracy against Constantine," he noted on 16th December, "and four of the generals who were killed perished in his anteroom in defending him. With the smallest beginning, however, nothing is more probable than a general rising in Poland; and what between that, Belgians and Piedmont which is threatened with a revolution the Continent is in such an imbroglio as this cannot be got right without a war; such a flame can only be quenched by blood." [71]

Lord Aberdeen noticed the embarrassment of Princess Lieven who, however, tried to spread the opinion that the rising would end with the incorporation of Poland within the Empire. In answer to this Wellington observed that the war in Poland would be more difficult for Russia than Nicholas seemed to imagine, especially if the Lithuanian corps – the withdrawal of which had been reported by the newspapers – were not employed. "If this is the case," wrote the Duke, "I don't think that the Emperor can now bring forward 70.000 Russians into Poland. That number will not be sufficient. However, I wish that he may be able to settle this affair." [72]

A similar wish was officially expressed by Palmerston. H.M. Government could give no instructions to be communicated to Russia "upon this delicate subject." However, "they cannot contemplate without shuddering the possible consequences of a war of extermination to be waged between the two nations". The British Ambassador might perhaps be able to help if an arrangement was likely between the two parties. In a private letter the Foreign Secretary requested Heytesbury to watch the conformity of any development of the Russian administration in Poland with the Vienna stipulations. The British Government did not intend to intervene in the contest. Things might be different if it appeared

[71] Cf. de Guichen, p. 245; *Greville Diary*, ed. H. Reeve, II, 98. Only one of the generals was killed at the palace of the Grand Duke; the others several hours later, in the streets and under quite different circumstances.

[72] Aberdeen to Wellington, 30 dec., Wellington to Aberdeen, 1 Jan., Wellington, *Supplementary Dispatches*, 1878, VII, 386, 389.

that Nicholas' own forces were unable to cope with the threat. "But the case is not ripe for such a course at present", wrote Palmerston. A few days later, on receipt of the first dispatches from St. Petersburg after the Polish news, the Foreign Secretary thought that leaving discretion to Heytesbury as to the time and manner of some friendly advice to be given by the British Ambassador to the Imperial Government with regard to a settlement in Poland appeared to have been a sound course.[73]

Grey also tried to employ Princess Lieven's influence for the same purpose; a war of extermination in Poland, which seemed only too likely, would both arouse the sympathy of all the nations for the Poles, and, possibly, strike a blow at Russian power elsewhere. "These, believe me, are not imaginary dangers," he continued, "and I earnestly hope that some means may be found of appeasing the storm which appears to me to be so threatening." Unfortunately, the Princess not only held a different view on the subject, but conveyed abroad the impression that the British Prime Minister "wished for immediate suppression of the Polish insurrection", much to the annoyance of Grey.[74]

In the meantime, in accordance with Sebastiani's advice, Prince Leon Sapieha came to London and went to see Grey. As he had a passport issued by Pozzo, and even dined at the Russian Embassy, Grey made no difficulty in receiving him, not expecting censure from Princess Lieven. Nor did he hold out to the Pole any views which departed from those expressed in his letters to the Princess. Reminded of his pamphlet on Poland, Grey said he had not changed in principle; but in view of the existing danger that the French would be driven by public opinion to attempt to annex Belgium, Britain must have an ally capable of counter-acting such a move. This could only be Russia. Less discouraging was Palmerston's statement, but, noted the Polish agent, it amounted to the same thing. For the rest, no better prospects were held out by other people. Talleyrand was reported to have said that the British Ministers did not want to listen to any proposals concerning Poland. The Duchesse de Dino's comment was: "La question polonaise qui

[73] Palmerston to Heytesbury, 31 Dec., F.O. Russia, and priv., Heytesbury Papers; Palmerston to Grey, 6 Jan., Howick MSS; cf. above, p. 63.

[74] Grey to Mme Lieven, 29 Dec., 5 Jan., Mme Lieven to Grey, 6 Jan., *The Lieven–Grey Correspondence,* II, 130-133. The letters of December 29 and 30 contained references to the alleged offensive words of Nicholas to Bourgoing, cf. above, p. 64.

rencontre une si juste sympathie sur le continent civilisé, ici n'inspire rien. Mme Lieven qui domine Lord Grey se sert de tout son crédit pour détacher à cet égard le cabinet anglais." More influenced perhaps by this opinion than by what he had heard from Grey, Sapieha thought that it would be good enough if Britain were not to oppose a French intervention.[75]

Princess Lieven was, however, far from content with the opinions prevailing in London with regard to Poland, even before she learnt of the instructions on that subject sent to Heytesbury. Yet, apart from the Press, there did not seem to be anything in London to take the place of the kind of speech Lafayette made in the French Chamber, where he said that, supported by the general wish in all Europe, Poland was about to become independent. The most characteristic pronouncement on the subject of Poland must be considered to be the comment which that undoubted friend of the Poles, Lord Holland, delivered on Lafayette's speech: "I wish Poland could be pacified & arranged without great military movements," he wrote to Grey, "though I also wish Lafayette had refrained from talking about it." [76]

About this time it became known through various channels that the Grand Duke Constantine had left all his archives at Warsaw; the Polish Government seized them but, contrary to Heytesbury's expectation, they were not published and only some of the documents were communicated abroad. Talleyrand, too, got hold of the news and caused Princess Lieven to make the following report. "M. de Talleyrand a dit hier à quelqu'un... que le soi-disant Gouvernement de Varsovie avait transmis à Paris une lettre de l'Empereur au Grand Duc Constantin, ..., par laquelle il apporte que l'Empereur avait pris la ferme résolution de faire la guerre à la France – cette lettre renferme des détails; il y est fait mention entre autres de 200 canons, etc..." And the French Ambassador was said to have commented that the Poles must be helped, "car si l'Empereur triomphe d'eux, nous aurions notre tour." [77]

[75] Sapieha, *Memoirs*, p. 115-16; Mme de Dino to Louis Philippe, 8 Jan., *Revue des Deux Mondes*, (1910) LVI, 336.

[76] Mme Lieven to Alex. Benckendorff, 2, 6 Jan.: "Vous concevez qu'ici, comme en France, la révolte des Polonais trouve des sympathies et des adhésions. Les journaux sont détestables sur ce sujet." Lie MSS. Cf. de Guichen, p. 245; Holland to Grey, 20 Dec., Howick MSS.

[77] Heytesbury to Palmerston, 20 Dec., Heytesbury Papers; Mme Lieven to Alex. Benckendorff, 7 Jan., Lie MSS.

That news was, at any rate for the British Government out-of-date and a different, more recent piece of news arrested their attention. In the second fortnight of December an account of Metternich's interview with the French Ambassador in Vienna circulated in the European capitals. The Austrian Chancellor was said to have regretted the error of Kaunitz in helping towards the partition of Poland and thus establishing a common frontier between Russia and Austria, about which the latter felt by no means comfortable. Other reports, too, confirmed not only that France was uneasy about the coming advance of the Russian forces into Poland, fearing that Nicholas' army, after a speedy crushing of the Polish resistance, would be tempted to cross the frontiers of the Empire, but also that Austria had speeded up the pace of her re-arming "à cause de la révolution polonaise dont on ne peut prévoir les suites..."; the Russian forces, collected for the purpose of suppressing the Polish revolt, arrived in the neighbourhood of the Austrian territories. In their turn, 60.000 Austrian soldiers were sent to guard the frontiers of Galicia, the Polish province of Austria. In Vienna "on tient conseil sur conseil [the subject being the cost of troops]", reported the Dutch representative. Meanwhile via diplomatic channels Metternich's pronouncement had become common knowledge in governmental circles. Lord Stuart de Rothesay reported it from Paris; Lord Holland wondered at it, and it no doubt influenced his view of the rôle of Austria as he later expounded it to the Poles as well as to his colleagues. Bülow found it necessary to mention it as late as May, 1831, and Palmerston told Princess Lieven about it early in January, whereupon she promptly wrote to St. Petersburg.[78]

On the other hand Palmerston made no great scruples about seeing Marquis Wielopolski, the official envoy of the Polish Govern-

[78] Maison to Sebastiani, 28 Dec., Sebastiani to Maison, 29 Dec., A.é. Autriche; Van Spaen to Verstolk, 30 Dec., R.A.; Stuart to Palmerston, 17 Dec., F.O. France; Holland to Grey, 23 Dec., Howick MSS; Mme Lieven to Alex. Benckendorff, 6 Jan., :"Ld Palmerston m'a confié un propos de Metternich dont les rapports officiels lui rendent compte. Il a dit sur les événemens de Pologne que le Pce Kaunitz avait fait une faute capitale en souffrant le partage de la Pologne, et que lui, Metternich, en reconnaissait tellement l'importance qu'il serait prêt aujourd'hui à céder la Galicie pour former une Pologne indépendante, barrière nécessaire à l'Europe." Lie MSS; A. Stern, *Geschichte Europas,* 1921, IV, 143, refers to Bülow's report as "Unbewiesen"; Stern's quotation, also in French, varies from the above which may be Maison's – I was unable to find it in A.é. Autriche. Bülow's account may be his own résumé.

ment. Wielopolski's arrival in London (delayed, as he had been refused a visa by Lord Stuart and only obtained one after producing a French passport, readily supplied by Sebastiani) received considerable publicity through the newspapers, and drew the attention of Greville, who noted it on 12th January.[79] The Foreign Secretary told the Polish envoy that Lieven had been informed of their meeting, nevertheless it would be better to arrange for a private interview. He then saw Wielopolski again on 24th or 25th January. Both interviews were limited to an exchange of information about the constitution of the Kingdom of Poland and the British standpoint in this regard, which did not go beyond what had already officially been conveyed to the Ambassador in St. Petersburg. The Polish envoy gave no account of the place of the meeting, but Van Zuylen's report of the end of January indicates that no particular secrecy seems to have been observed. "The Foreign Office presented on the 25th a peculiar sight," wrote the Dutch plenipotentiary. "Upstairs the Conference have been sitting, including ourselves [i.e., Falck and Van Zuylen]. Below, Van de Weyer and Vilain XIV were waiting; in another room the Ghent deputation [the Orangists] and in still another room the Polish representative of the Insurrection!!!"[80]

The mere presence of Wielopolski in London was in itself a problem to the Russian mission. Tirelessly watching the slightest appearances in favour of the Poles Princess Lieven had already before 8th January enquired of Palmerston if he had had any communications with the Poles. He answered that he had not – though he had seen Sapieha and must have, by that time, arranged for the first interview with Wielopolski – but that if he should have any, all

[79] Wielopolski himself requested the newspapers to announce the arrival of an envoy from the National Government of Poland. Palmerston, therefore, at first categorically refused Sapieha's request to let him introduce Wielopolski. Sapieha, *Memoirs*, p. 119; *Greville Diary*, II, 107, refers to the arrival of an envoy from the Poles which "had naturally given umbrage to the Lievens." Hobhouse described a public dinner given in honour of Wielopolski, at which, to his great annoyance, he took the chair. "The Polish Marquis spoke for more than one hour in French. He was in bad spirits and told me that, although he had letters of introduction to several persons connected with the government, not one of them would receive him. Prince Czartoryski had given him a letter to our Lord Chancellor [Brougham] who behaved to him like the rest." Hobhouse, *Some Account of a Long Life*, II, 84-5. The dinner was probably that held in March, cf below, p. 119.

[80] H. Lisicki, *Aleksander Hrabia Wielopolski, Margrabia Myszkowski*, II, 343-352; Van Zuylen to Verstolk, 28 Jan., R.A.

would be arranged in a spirit of friendliness to the Emperor; and that meant, judging from the rest of the letter, that by restoring to the Poles the constitution granted to them by the Emperor Alexander, Nicholas would acquire "plus de gloire et de vrai pouvoir qu'en roulant sous ses pieds les cadavres de cinquante mille Polonais." But Palmerston admittted that he did not expect that his remarks would encounter a favourable reception.[81]

Grey, according to Wielopolski, did not turn up for the meeting privately arranged on 8th January, and this the Polish envoy ascribed to a very important and very drastic conference on Belgian affairs, held one or two days before; but some days later Princess Lieven noted the presence of Wielopolski at Talleyrand's, on which occasion "M. Esterhazy a causé avec le député, il me l'a dit lui-même, Bülow *prétend* ne l'avoir pas fait." She learnt the fact from Grey, who again was at great pains to explain to her the difficulty of remaing silent upon the affairs of Poland. The Princess continued, however, to tell him "que le conseil le plus voilé ou le plus délicat" would invariably meet "une réponse sèche et désobligeante." The French Embassy was soon closed to Wielopolski, on account of Russian influence as he believed.[82]

In Paris, however, the Ministers were not in a position to avoid contact with the Poles. During the two meetings which Sebastiani had with another Polish agent, Wolicki, the Minister held out no promises but only expressed general sympathy, whilst on the second occasion, 8th January, he coupled this sympathy with a vague utterance of the possibility of a "more decisive attitude" of France towards the Polish movement if it should hold out beyond 1st March.[83] Meanwhile the credentials of Pozzo had arrived, and the departure of the Duc de Mortemart was decided upon; the Ambassador declared to Wolicki: "Je n'ai véritablement accepté cette pénible ambassade que par sympathie pour votre illustre nation et dans

[81] Palmerston to Mme Lieven, 8 Jan., Lie MSS. The concluding passage of the letter reads: "Eh, mon Dieu, que vous êtes impertinent, me direz-vous, verrez-vous, Monsieur, à vos propres affaires dont il me paraît que vous avez déjà les mains remplies. Eh bien, soit – mais du moins je vous ai dit ma pensée."

[82] Lisicki, III, 343 and I, 23; Mme Lieven to Alex. Benckendorff, 22 Jan. (Italics mine), 1 Febr., Lie MSS; Grey to Mme Lieven, 24 Jan.: "If you look at "The Times" you will be able to judge of the course public opinion here is likely to take if the business is protracted, and of the difficulties which may ensue." *The Lieven–Grey Correspondence,* II, 144.

[83] J. Dutkiewicz, *Polska a Francja w r. 1831* (Poland and France in 1831), 1950, p. 58.

l'espoir d'arriver à temps auprès de votre roi constitutionnel pour vous réconcilier avec lui." Mortemart may have been sincere in making this statement– it was known that he only reluctantly associated himself with the Orleanist régime, while he did not expect that this statement would be published by a French newspaper; on 26th January, at the time of the publication, he had just had his secret interview with a Polish envoy on his way through Prussian territory and the dethronement of the Romanovs had already been pronounced by the Polish Diet.[84]

Nicholas may have hoped for a peaceful settlement with the Poles – on his own terms – but the feeling in St. Petersburg hardly allowed for a settlement as a result of French mediation. Already at the time of the French declaration against any Russian intervention Nicholas gave the French *Chargé d'affaires* an account of the intercepted Polish letter which he considered as evidence of the Polish revolution having been prearranged with France, while Bourgoing was convinced that owing to Russian military superiority "la question polonaise est dès à présent jugée." This also accounted for Nicholas' attitude in Belgian affairs: in an audience early in January he seemed satisfied with France's pacific declarations, but the idea of armed intervention in Belgium in favour of the King of the Netherlands did not leave his thoughts; when Bourgoing represented to him the international dangers of such a solution, Nicholas answered that exaggerated cautiousness accounted for the loss of Belgium by William I. In a subsequent dispatch, concerning African affairs, Bourgoing said that in view of the preoccupation of the Emperor with Poland and Belgium Mortemart would experience no difficulty in obtaining Russian consent to the recognition of the conquests in Algiers.[85]

As far as Polish affairs were concerned, Mortemart could not expect any success. Nicholas' distrust of France received a fresh stimulus at the reception of some news which had very little foundation, namely, that as soon as the outbreak in Warsaw became known, a great number of French officers "s'est mis en route pour se ranger sous le drapeau de la révolte." As Sebastiani ostensibly kept aloof from the Polish movement, while French military preparations continued and would involve the necessity of keeping those of-

[84] Barosz, "La révolution polonaise de 1830 et la déchéance du tzar Nicolas Ier", *Annales de l'école des sciences politiques*, 1895, pp. 426-7; cf. Dutkiewicz, p. 62.

[85] Bourgoing to Sebastiani, 23 Dec., 1, 4, 5 Jan., A.é. Russie. Cf. above p. 60.

ficers at home, St. Petersburg circles considered this news as evidence of an organised secret measure of the French Government. Heytesbury, who had just received his instructions with regard to Poland, wrote back to say that in his opinion no interference with regard to Poland would meet with a favourable reception.[86]

In the middle of January the news of *the* Belgian Protocol, No. 7, began to arrive. The Protocol was found unsatisfactory because it contained no statement with regard to the rights of the House of Nassau; dispatches from The Hague tended to strengthen this view. On 29th January Nesselrode told the French *Chargé d'affaires* that governments should not yield to the Belgian revolutionaries who persisted in their refusal of the Prince of Orange. Bourgoing then again represented the dangers inherent in all forcible measures; could not the King of the Netherlands be advised by the Emperor to make some sacrifices in the cause of peace? The answer was quite positive: "Vous pouvez être persuadé que ce ne serait jamais de sa part que viendront de semblables conseils", said Nesselrode. Bourgoing thought he had obtained some success in having the Vice-Chancellor agree to one of the most delicate points, that an armed intervention in Luxemburg by the Federal troops should take place "aussi tard que possible". Heytesbury, however, considered his tone about Luxemburg had changed a good deal in that France now admitted the possibility of a military action. In the meantime, the basic attitude of the Imperial Government remained unaltered: Nesselrode expressed his dissatisfaction that the Conference applied the same tone both to William I and to the Belgian Provisional Government. Lieven's signature on the Protocol of 20th December had only been approved because it still left open the only safe solution for Belgium, the placing of its sovereignty with a prince of the House of Nassau.[87]

Some days later new reports from Lieven seemed to supply new evidence of French duplicity in Belgian affairs as well as in their pacific assurances. Sebastiani was said to have seen Pozzo on 22th January, and after a review of the Belgian difficulties – the weakness of the Orange party, and the French complaints with regard to the injurious treaties of 1814/15 – the Minister proposed a partition of

[86] Heytesbury to Palmerston, 21, 22 Jan., F.O. Russia; Van Heeckeren to Verstolk, 23 Jan., R.A.

[87] Heytesbury to Palmerston, 11, 19, 31 Jan., F.O. Russia; Bourgoing to Sebastiani, 29 Jan., A.é. Russie.

Belgium to the Russian Ambassador. To the further amazement of Pozzo di Borgo Sebastiani supported his proposal by adding "qu'en cas de guerre... la France saurait trouver des auxiliaires dans toute l'Europe, qu'il dépendait d'elle de fomenter des troubles partout où elle voudrait; qu'on est informé à Paris que toute l'armée prussienne est gangrenée..." etc. Lord Heytesbury considered this to be just another example of France's taking advantage of the troubled state of Europe, quoting "the occupation given to the Russian armies by the Polish insurrection" as its first instance. The bad faith of the French Government was proved beyond any doubt, thought the Netherlands Minister; "L'Empereur est content de le voir jeter le masque", he wrote.[88]

A few days earlier the Netherlands Minister, having completed his record of the protocols of the Conference up to 9th January, and with adequate instructions from home, went to see the Vice-Chancellor. "Le Cte Nesselrode a commencé par me témoigner combien il était désespéré que les Plénipotentiaires russes se fussent regardés comme obligés de signer le protocol de 20 Xbre", reported van Heeckeren. His reply was that he only understood that the plenipotentiaries had not been disavowed entirely, because there was no question at all of the rights of the House of Nassau in the Protocol; but consent to the full independence of Belgium – confirmed by new instructions to Lieven sent at the end of January – after all, gave evidence of yielding to the revolution. Nesselrode defended himself saying that these instructions would only be carried out if the King agreed; and he was sure that the King would. But Van Heeckeren still protested – the arrangement proved that Russia had given way. In reply, the Russian Minister made the following statement:

"Je l'atteste devant Dieu..., la Russie fait tout ce qu'elle peut pour défendre les intérêts du Roi des Pays Bas; mais accordez aussi la part des événemens: de quelque coté que nous tournions nos regards, que voyons-nous? La France hostile, l'Angleterre plus qu'incertaine et sans force, la Prusse sans ministre..., l'Autriche

[88] Heytesbury to Palmerston, 8 Febr., F.O. Russia; Van Heeckeren to Verstolk, 9 Febr., R.A. There is no mention of this interview in A.é.; the account given by de Guichen, p. 280, from Pozzo's dispatch of 13/25 January does not suggest that any incident did take place. It is possible, therefore, that this was an instance of a false alarm, similar to the case of the alleged offensive words of Nicholas to Bourgoing. For the French plans for the partition of Belgium at that time see below, p. 108.

faible et se taisant, nous-mêmes les affaires de la Pologne sur les bras et de la sorte en présence des immenses événemens; mais laissez-nous sortir des embarras que nous suscite l'insurrection de Varsovie et vous voyez si la Russie saura reprendre le langage qu'elle doit tenir dans les affaires de l'Europe." [89]

The weeks past had already witnessed a certain reassurance on the part of the Imperial Governement, while at that time the end of the Polish revolt was expected, any day and with mixed feelings, in most European capitals.

[89] Van Heeckeren to Verstolk, 5 Febr., R.A.

CHAPTER III

NEMOURS VERSUS ORANGE AND THE BATTLE OF GROCHOW

If the outbreak of the rising in Warsaw complicated affairs in Europe, already unsettled by events in Paris and Brussels during the previous summer, the early months of the year 1831 brought further aggravation of the international situation. A breakdown of the Polish movement would confront Europe with large Russian forces, ready to march, and, it was felt, likely to be employed in Belgium or elsewhere in the West; a possible intervention by foreign Powers in Poland, such as to exclude military operations, and, consequently, a full Russian mobilization, was uncompromisingly handicapped not only by the unwillingness of St. Petersburg but also by the dethronement of Nicholas, pronounced by the Polish Diet. At the same time difficulties had arisen in Belgium about the election of a king. Popular feeling seemed to favour either the Duc de Nemours, a son of Louis Philippe, or the Duc de Leuchtenberg, the stepson of Napoleon, neither of whom was acceptable to both of the Powers most interested, France and Britain; there existed a party for the Prince of Orange, whose candidature enjoyed the special support of Russia and was greatly assisted by the British member of the Conference's representation in Brussels, Lord Ponsonby. Thirdly, in the latter half of February commotions began in Italy and menaced Austro-French relations. Placed in the midst of affairs so complicated the French Ambassador in Vienna, Marshal Maison, thought that war was inevitable. In this spirit he wrote privately to his colleague at Constantinople; and his opinions made General Guilleminot act in a way which caused a major diplomatic incident several weeks later, with repercussions in London as well as in St. Petersburg.[1]

I

Almost at the same time as the French Government tried to raise

[1] Maison to Guilleminot, 24 Febr., 1831, A.é. Autriche. See further below, p. 143.

apprehensions in London with regard to the consequences of a Russian victory in Poland, Poland was the subject of concern and mixed feelings on the part of the other partners of Nicholas. Lord Cowley, the British Ambassador to Vienna, began his dispatch of 7th January by a review of opinions on the effect of military operations on the Vistula. Even a prolonged resistance by the Poles would give rise to serious disturbances in Germany and Italy, encouraging the people and dispersing the forces of the two great German Powers guarding the Eastern frontiers. On the other hand, Russian success, which would bring an army of 150.000 along the frontiers of Germany, would make the Confederation adopt much firmer language towards France. This success could, however, be achieved in a variety of ways. The Austrian Chancellor informed Cowley of his refusal of the secret offer of the Polish crown for the Archduke Charles; this refusal was coupled with the advice to the Poles to submit. Metternich also said to Cowley that the Dictator Chlopicki was seeking to come to terms with Emperor Nicholas, and this, the Ambassador found out, was the opinion of many Poles in Vienna. Yet, a week later, he reported the Austrian view from St. Petersburg – that several months would have to elapse before Poland could be pacified.[2]

The Polish rising caused even more inconvenience to the Prussian Government. The French *Chargé d'affaires* in Berlin, Mortier, was able to ascertain to his surprise that a gentleman from Prussian Poland, Count Raczyński, who had an established position in Berlin, had approached several persons in Court and Government circles about the possibility of mediation by Frederick William III between Nicholas and his Polish subjects; to the even greater surprise of Mortier Ancillon denied Raczyński's overtures, of which the French *Chargé d'affaires* knew the Russian Ambassador in Berlin had learnt from the Prussian Government. It was known in diplomatic circles that a senior officer, sent by the Government to Vienna under the pretext of presenting the King of Hungary with the insignia of the Order of the Black Eagle was really instructed to establish a common line of antagonistic policy towards Poland between the two Governments. On the other hand it appeared to Mortier that the King of Prussia was almost anxious to write to his Imperial son-in-law upon the subject of mediation if the Poles asked for it.[3]

[2] Cowley to Palmerston, 7, 15 Jan., F.O. Austria.

[3] Mortier to Sebastiani, 11 Jan.: "V.E. sera peut-être étonné qu'un sujet du Roi de Prusse intervienne auprès de son Gouvernement en faveur du parti qui

The interest of France in the Polish question continued, though in a different way. The French proposal made to Britain through Lord Stuart de Rothesay for a joint note to St. Petersburg to establish that there would be no change in the status of the Kingdom of Poland as set-up in 1815 was refused; Sebastiani, however, took the opportunity to stress to Lord Granville, the new Ambassador, that the French Government appreciated the British standpoint that an incorporation of Poland with Russia would be inconsistent with the Treaty of Vienna. In his reply to Mortier Sebastiani approved his attitude towards Raczyński but pointed out that all Polish agents should be well received because of the interest and sympathy which their cause aroused in France; the French Government, if unable to act in favour of the Poles, had to reckon with public opinion which demanded such an action.[4]

Though the French proposal had been refused, in England, too, some uneasiness was felt about the future status of the Kingdom of Poland in consequence of the approaching struggle. Princess Lieven reported with regret that instructions concerning Poland had been sent to Lord Heytesbury; this came to her knowledge the day after she had noted that the interest of Government circles in the contest on the Vistula did not stretch beyond sympathy; it was through her that Grey advised the observance of the Treaty of Vienna by no less a remark than that in 1815 a mere suggestion of the incorporation of Poland in Russia "very nearly produced war." [5]

If only through newspapers, the concern of Paris and London with the Polish question could not pass unnoticed in Berlin, and the Prussian Government thought it advisable to calm down what appeared to be public excitement. Early in February Ancillon assured the British Minister that the Emperor Nicholas had no intention of incorporating Poland in Russia, and as apprehensions were

s'est emparé du pouvoir à Varsovie, et que des Princes de sa maison et des hommes influens de son Gouvernement prêtent oreille à ses ouvertures, surtout lorsqu'il n'a pas un caractère officiel. Le fait peut paraître étonnant au premier abord, mais, si d'un côté, on se rend compte des embarras que causent à la Prusse les événemens de Pologne et de son désir qu'ils ayent promptement un fin, cependant elle est disposée à recevoir toute communication qui annoncerait de la part du dictateur la volonté de réclamer sa médiation." On the real purpose of General Roeder's mission to Vienna: Cowley to Palmerston, 20 Jan., F.O. Austria.

[4] Granville to Palmerston, 7, 21 Jan., F.O. France; Sebastiani to Mortier, 1 Febr., A.é. Prusse; cf above, p. 87.

[5] Mme Lieven to Alex. Benckendorff, 1, 2 Febr., Lie MSS; Grey to Mme Lieven, 27 Jan., *The Lieven-Grey Correspondence,* II, 148-9.

entertained in France and England about the fate of that country, he would instruct Bülow to make an appropriate statement on the subject to Palmerston. Notwithstanding this assurance Chad took another opportunity of speaking on the subject, which was supplied by the news of a proclamation addressed to the Poles by Nicholas. In an interview with the British Minister Bernstorff said that this proclamation amounted to the announcement of great changes in the governmental system in Poland. To this Chad replied that an incorporation within the Empire would play in the hands of France, who was looking for a pretext for war. The Prussian Minister did not directly deny this suggestion; he merely observed that France, professing the idea of self-determination of peoples, would be compelled to admit that the Poles had dissolved the Treaty obligations by their rising. Apparently, though Chad could not readily confirm this, Berlin was confident of the speedy suppression of the revolt; it was even supposed that after the defeat the Polish cavalry might attempt to force their way through Germany to France.[6]

Similar expectations were held in the other two Eastern capitals; by the second half of February there was a general belief that the Russian forces then entering the Kingdom would soon bring an end to the Polish movement opening the way for all that had been feared in the West. "Toutes chances de succès semblent être du côté des Russes", wrote Bourgoing from St. Petersburg. French reports from Berlin, like those of Chad, were not so alarming, but both Maison and Cowley noted the constant increase of warlike tendencies around the Russian Embassy in Vienna.[7]

It seemed obvious that France had to do something to avert the apparent Russian menace, directed – and there were no illusions as to this in Paris, nor elsewhere – against almost everything that the July revolution stood for. The mission of the Duc de Mortemart, planned for weeks, was designed to serve this purpose; his instructions give ample evidence of the great concern of the French Government with regard to Mortemart's success. His mission, it was said, had for its only object the strengthening of the friendship between the

[6] Chad to Palmerston, 6, 16, 20, 23 Febr.; the report of the 16th reads: "The Russians here, tho' quite confident of victory, are not without anxiety as to the price at which it will be obtained." F.O. Prussia.

[7] Cowley to Palmerston, 8, 16 Febr., F.O. Austria; Cowley's own opinion, however, was that a considerable number of Russian troops would have to be left in Poland. Bourgoing to Sebastiani, 28 Jan., Mortier to Sebastiani, 15, 20 Febr., Maison to Sebastiani, 16, 26 Febr., A.é. Russie, Prusse, Autriche.

two monarchs and the two nations, "qui peuvent le plus influer sur l'avenir de l'Europe et qui sont le plus en état de lui assurer le bienfait de maintien de la paix. Cette mission est donc pour ainsi dire, Européenne", proceeds the document, "et en effet, l'Europe entière y est intéressée: elle en suivra les phases avec sollicitude." The background of the European situation is then immediately developed. "Deux questions principales et d'une nature également grave dans leur principe comme dans leurs conséquences fixent en ce moment l'attention des hommes d'état et excitent l'intérêt des peuples: je veux parler de la révolution de Belgique et de l'insurrection de Pologne." The exposé of French views concerning Belgium with great care aims at showing all the good will possible. The French Government was aware of the Belgian inclination for a reunion with France; it repulsed the idea entirely, yet was mindful of the influence which the state of affairs in Belgium exercised upon France. The Emperor of Russia should, therefore, appreciate the loyalty of the King of the French, who in these circumstances was determined to refuse the offer of the Belgian crown for his son. Events in Poland would not change French policy in Belgium; France wanted for Poland a reconciliation with the Emperor and peace. "Si nous pouvons contribuer à ce résultat par quelques démarches, nous les ferions avec empressement; mais cette intervention tout amicale, toute pacifique de notre part ne saura avoir lieu qu'autant que la Russie non seulement y consentirait mais encore paraîtrait la désirer." [8]

At the time of writing the above Sebastiani expressed a belief that Nicholas was disposed to negotiate – though he would not make the first step – as otherwise a victory "would be dearly bought", but later reports show that the French Government could not very well expect a welcome from Nicholas to any interference on its part, while it was equally aware of mixed, if not depressed feelings at Warsaw.[9]

The Poles would have been surprised to read Mortemart's instructions; they hoped much from his mission. Accordingly, a Government official had been sent to meet him secretly on his way to St. Petersburg; the interview took place on 24th January in a stage coach near the Eastern frontier of the Grand Duchy of Posen, and lasted

[8] Sebastiani to Mortemart, 9 Jan., A.é. Russie. Mortemart left Paris on the 10th; the credentials of Pozzo were presented on the 8th; cf. above p. 88 and 94.

[9] Granville to Palmerston, 7, 21 Jan., F.O. France; Durand to Sebastiani, 18 Jan., A.é. Pologne; cf. below, p. 105.

three hours. Mortemart listened to the account of the rising, and himself spoke much about the situation in France and elsewhere, which he painted in dark colours; he advised further negotiations with Nicholas; he promised no help from France in the struggle. The only hopeful thing he said was that France would prevent Prussia from interfering. The Polish agent, quite disappointed, said that the Poles were fighting for their independence and would either obtain this or nothing; to this the French Ambassador replied: "Eh bien, je vous le dis avec douleur mais avec une profonde conviction: ce sera rien." [10] Yet he found the military aspect most important; closing his report he observed that the Poles could have gone unopposed as far as Dvina had they prevented in time, as they could easily have done, the joining of the forces of Diebitsch with those of the Grand Duke Constantine.[11]

Sebastiani approved of Mortemart's attitude towards the Polish agent; he wrote in reply that a terrible struggle, which would end with a Russian victory, was to be expected, and instructed the Ambassador to seek mercy from the Emperor for the defeated.[12]

France was also envisaging other consequence of the approaching Russian campaign in Poland; great concern had been shown in Paris at the first news of the Russian mobilization and, as has been pointed out above, the reaction in St. Petersburg, in consequence of the French protest could not but greatly increase the apprehensions entertained in the French capital. London was the first to be approached on this subject; in the interview of 20th January, already referred to, Sebastiani pointed out to the British Ambassador that, if a Russian victory was rapid and decisive, "large Russian forces in Poland would present an attitude so menacing, as greatly to increase the probablity of war." [13] Subsequently, this matter was raised in Vienna, now greatly embarrassed with the state of Italy, and in Berlin, always more sensitive as well as more vulnerable. This time, however, even less success was achieved in the Prussian capital.

[10] J. Dutkiewicz, *Poland and France in 1831,* pp. 62-3, cf. St. Barzykowski, *Historya Powstania Listopadowego* (The History of the November Rising), Poznań 1883–4, II, 66 sq; quotation from P. Thureau-Dangin, *Histoire de la monarchie de juillet,* Paris 1884–92, I, 168.

[11] Mortemart to Sebastiani, 28 Jan. (from Königsberg), A.é. Russie. The meeting of Mortemart with the Polish agent was no secret in Berlin, not even to the diplomatic corps, Chad to Palmerston, 30 Jan., F.O. Prussia, cf. Hillebrand, *Geschichte Frankreichs,* I, 169.

[12] Sebastiani to Mortemart, 20, 28 Febr., A.é. Russie.

[13] Granville to Palmerston, 21 Jan., F.O. France.

Metternich refused to consider disarmament as long as the insurrection lasted in Poland; but he assured the French Ambassador that nothing was to be feared from Russia, as she would have to lead her armies through Prussian and Austrian lands, and the respective governments would not allow her to do so. But Maison observed at the same time: "Le langage de la Russie est on ne peut pas plus belliqueux." On a subsequent occasion Metternich admitted that some uneasiness might have been felt about Russian intentions after their forces arrived in the very centre of Europe, but at this time he felt none; yet Maison again was compelled to record warlike sentiments around the Russian Embassy.[14]

The French *Chargé d'affaires* in Berlin was glad to report that Ancillon changed his tone with regard to Belgium; hopes and fears on account of the Prince of Orange's chances were reduced to "des regrets impuissants"; but from the interview with Bernstorff Mortier drew the conclusion that the withdrawal of the Duc de Nemours had been, in the opinion of the Prussian Government, dictated by fear. Nor was the King of Prussia inclined to depart from his wish to maintain the House of Nassau in Belgium; he did, however, call the Russian mobilization "une grande imprudence" and affirmed that the Russian army would not be allowed to pass through his dominions. Apart from this, military circles hoped for a Russian victory as a check against the war party in the French Cabinet, but realized the difficulties of the military operations.[15]

In the meantime the Duc de Mortemart completed his journey, which took more than four weeks, and on 8th February arrived in the Russian capital. The audience which on 12th February Nicholas granted to the French Ambassador was almost entirely devoted to Polish affairs. "S. M. m'a, d'abord, remercié des dispositions que le Gouvernement du Roi apportait dans la question de la Pologne," reported the Ambassador, "ainsi que des efforts faits par l'administration pour neutraliser le fâcheux effet, qu'auraient pu produire sur la population Polonaise, déjà si exaltée, les excitations

[14] Maison to Sebastiani, 15, 16, 25, 26 Febr., A.é. Autriche.

[15] Mortier to Sebastiani, 15, 19, 20 Febr., A.é. Prusse; Gen. Wolzogen to Natzmer, 17 Febr.: "Dass Soult und Sebastiani den Krieg wollen, bin ich auch überzeugt... Hoffentlich werden die Russen bald im Besitz von Warschau sein, obwohl die nun mit Eis gehende Weichsel viele Schwierigkeiten legen wird. Ist die polnische Angelegenheit beseitigt, so werden 150,000 Mann an unsren Grenzen disponibel." G. E. v. Natzmer, *Denkwürdigkeiten aus dem Leben des Gen. Oldwig von Natzmer,* Berlin 1887–9, I, 262.

contenues dans nos feuilles publiques." Mortemart only tried to strengthen this impression of the Emperor. Nicholas spoke of the dethronement of his family, recently (on 25th January) pronounced by the Polish Diet, "avec plus de regret et de pitié que d'indignation", so it seemed to the Ambassador. The few sentences concerning Belgian affairs led him to the conclusion that the Emperor admitted there was but little chance for the Prince of Orange and thought, in any case, that this matter might rest with the Russian plenipotentiaries at the Conference. Later, when St. Petersburg learnt of the refusal of the Duc de Nemours, Nicholas sent his principal aide-de-camp, Benckendorff, to Mortemart to express his approval. Unfortunately, the Ambassador had to supplement his report with information about renewed hopes of the Emperor for the Prince of Orange's return to Brussels.[16]

More detailed explanations, however, which Mortemart made on the ground of his later instructions, were coldly received. Nesselrode complained of the French refusal to accede to the Conference's decision with regard to the division of the debt and "de la manière dont les députés belges étaient reçus à Paris." Mortemart believed that his reply satisfied the Vice-Chancellor, "néanmoins, il n'était question d'aucune réponse, lorsque les nouvelles de Pologne sont arrivées." This was the news of the first encounter between the Russian and the Polish forces, the battle of Stoczek on 14th February which, though a comparatively minor engagement, was a complete Polish success. "Alors on m'annonce le départ subit d'un courrier", continued the French Ambassador, "pour remercier le Roi de ses démarches, mais, sans doute, pour une autre cause..." [17]

This occasion prompted Mortemart to describe at some length his position at the Russian Court. "Tout ici est contre nous: la diplomatie étrangère et les courtisans. On nous a reconnus par crainte et, maintenant, l'Empereur tiendra sa parole envers et contre tous. Quant à son amitié, nous ne l'acquerrons également que par la crainte; mais nous pourrons aussi sûrement y compter." Accordingly, Mortemart saw little chance to bring about a real *rapprochement* between France and Russia, and thought it better to withdraw at the first appearance of failure.

[16] Mortemart to Sebastiani, 14, 23 Feb., A.é. Russie; cf. also below, p. 110 sq.

[17] The debt was the subject of Protocol No. 9, of 27 January. Mortemart to Sebastiani, 28 Febr., A.é. Russie. cf. R. F. Leslie, *Polish Politics and the Revolution of 1830*, 1956, p. 166.

Like his report of the meeting with the Polish agent, the conclusion of his dispatch stated somewhat regretfully, the military position: "Les fautes du général Diebitsch sont patentes. Les Polonais n'ont su en profiter. La question est encore indécise; ils seront néanmoins écrasés." [18]

A rather singular comment on the first *démarches* of Mortemart was delivered by the Netherlands Minister in St. Petersburg; according to him the Ambassador denied having received any instructions in favour of the Poles; Mortemart himself does not record any statement to this effect in his account of the first audience with the Emperor. Baron Van Heeckeren found out, however, that Mortemart had been trying to enlist the support of the British Ambassador; speaking to Lord Heytesbury the French Ambassador had again referred to difficulties which would arise after the Polish defeat; the presence of considerable Russian forces in the very heart of Germany "changerait entièrement la face des choses et serait à la longue une position que ni ces deux puissances ni la France elle-même ne pourraient tolérer." Van Heeckeren's comment of course was that Russia would certainly profit from this circumstance to adopt an attitude more convenient for her policy "et nous verrons plus tard si effectivement le Duc [de Mortemart] est autorisé de faire des représentations sérieuses à cet égard." [19]

Thus it appeared at least as if the Netherlands Minister attributed to France the intention of influencing the political situation by diplomatic steps in favour of the defeated Poles.

The French Government, already in a difficult position with regard to Belgium, needed Mortemart's success badly. From Paris Granville was confidentially reporting repeated warnings from the Russian Ambassador of an imminent French attack. Eventually, it seemed to Granville that "Pozzo appears to wish to commit us to a quarrel with France," but that the Russian Ambassador's colleagues were less anxious to work in this direction; no doubt, thought Granville, because they would have to sustain a French attack first.[20]

[18] Mortemart to Sebastiani, 28 Febr., A.é. Russie.

[19] Van Heeckeren to Verstolk, 2 Mar., R.A.

[20] Granville to Palmerston, 9, 11 Febr., Palm MSS. The letter of the 9th reads: "Pozzo di Borgo has for some days been showing particular empressement to converse with me. His object means to be to convince me that Sebastiani is not to be believed, that the King, from his wish to be well with all parties, is not to be relied upon; that war is inevitable and all the Powers of Europe should

The British Abassador had no more success in trying to mitigate French re-armament. Already at the end of January Granville reported that a French levy of troops, previously fixed for April, was to take place immediately. In the beginning of February the question of Nemours' election disclosed, in Palmerston's words, "their [French] designs upon Belgium, and the underhand proceedings which they are carrying on with reference to that country. Yet, where nobody threatened them, their war preparations continued." On 22nd and 23rd February Granville devoted two interviews with Sebastiani to this problem. The Minister, however, declined to consider a suggestion of disarmament. As if he had been listening to conversations between Pozzo and the British Ambassador, he referred to the repeated attempts of Russia to create an anti-French coalition during the previous months. He mentioned a new levy of troops in Spain, and eventually, on both occasions, he recalled that the Court of Austria refused to reduce its armed forces whilst the state of Poland remained unsettled. Nevertheless, though by no means satisfied with French policy, Palmerston considered it advisable to tell Lieven that Pozzo's alarming reports were exaggerated.[21]

Soon, however, there was enough convincing evidence that there was another source of trouble as well; an anti-French coalition had not been a mere nightmare of Sebastiani. The information arrived from Vienna. Metternich had written to St. Petersburg to remind the Imperial Government of their promises to support Austria against France in Italy, where Hapsburg princes faced revolution. Now everything depended upon the pacification of Poland; British policy could be foreseen. In view of the French re-armament, which, however, would not be complete before April, it was imperative for the three Courts (Vienna, St. Petersburg and Berlin) to reach the fullest agreement before that time: after the suppression of the Polish revolution they would have to act without delay. The British Ambassador succeeded in secretly acquiring this dispatch, written in February, at the beginning of March. He had previously asked Metternich whether any change would take place, "whenever the affairs of Poland should be settled", and was given to understand

---

without delay by establishing a concert between them be prepared to resist the attack of France."

[21] Granville to Palmerston, 28 Jan., 24 Febr., F.O. France; Palmerston to Granville, 15 Febr., Bulwer, II, 41, cf. below, pp. 130-1; Palmerston to Lieven 25 Febr., Lie MSS.

that there would be none. He could not, however, help reporting "the rumours... still prevalent" in Vienna indicating the contrary; the acquisition of that confidential dispatch gave rise to his conclusion about the inevitability of a continental war.[22]

Whether with a view to attacking France, or being able to counter-attack, or simply to enable Russia to raise the standing of her diplomats, a victory in Poland could not but be impatiently looked forward to in St. Petersburg as well as in the Russian foreign missions. To the Belgian troubles the British Government had to add internal difficulties, and that, noted Mme Lieven about the middle of February, kept down their attitude towards France, and was as bad for them as for the other Powers in the Conference: "la France se gênera moins", she wrote, "et pour que Diebitsch tarde à frapper un grand coup, nous avons à avaler bien des insolences françaises et des poltronneries britanniques." Fortunately, she thought, it was just a question of time. "On admet ici que la Pologne est une affaire finie", commented the Princess some days later; "les rapports de Berlin le disent, tout le monde l'espère, personne n'ose le proclamer." [23] At the time of the writing the grand coup happened to be only a matter of days away.

II

At this time Anglo-French relations were growing more and more difficult on account of the need of a permanent solution for Belgium.

Count Flahaut, a friend of Talleyrand, but equally well acquainted with English Whig society arrived in London on 22nd January, ostensibly with no specified purpose, but alleged to have brought with him a plan for the partition of Belgium, an idea which in France seemed a likely one to solve all the problems around the revolution in the Low Countries. Apart from British views, even on formal grounds no consideration could be given to that project, as Flahaut arrived just after the signing of a Protocol establishing the neutrality of the Belgian territory which even more that the Protocol of 20th December became a milestone in the progress towards the independence of Belgium. Flahaut would of course not act in any but

[22] Cowley to Palmerston, 5, 16 Febr., 4 Mar., F.O. Austria.

[23] Mme Lieven to Alex. Benckendorff, 16 Febr., to Nesselrode, 20 Febr., Lie MSS.

a strictly confidential way, which was even more likely to give opportunity for comments. The Dutch plenipotentiaries observed that through the partition of Belgium the French Government had hoped to settle affairs with a view to preventing the Belgian Congress from proclaiming a reunion with France and also to being able to help the Poles. The latter intention could be considered self-evident from the popular feeling in France; and Lafayette even approached the British Ambassador to express the opinion that a common action by France and Britain could save the Kingdom of Poland from annihilation, which was threatening that country as a result of the approaching contest.[24]

The suggestion having once been officially refused, it was now again unlikely to attract the British Government, as the state of France and Belgium was far from promising assistance in the negotiations of the Conference. France feared the election of Leuchtenberg to the throne of Belgium – reported Granville extensively on the same day – who as a member of the Bonaparte family could not have been tolerated in the neighbourhood of Orleanist Paris; nor did she any longer profess support for the Prince of Orange, in the belief that this would mean civil war in Belgium. The French Government succeeded in convincing the British Ambassador that this was its only reason.[25]

But Grey, if only from his private correspondence with Flahaut, was entitled to hold different views on this question. On the eve of his arrival in England Flahaut wrote to him, pressing the question of the Belgian crown in such a way as to exclude all acceptable candidates, Leopold, "an English Prince", as well while the election of the Prince of Orange "would only prepare another revolution, and the reversion to France in two or three years." Yet an election must take place; "There is a revolutionary wind blowing", he wrote; hence all the five Governments of the Great Powers must keep the peace with one another, which, in Flahaut's opinion, also implied Nicholas' taking no offence at an offer of mediation made by Britain and France, and showing clemency to the Poles.

In his answer Grey mixed strong opinions with pacific assurances.

[24] Falck and Van Zuylen to Verstolk, 28 Jan., R.A.; Granville to Palmerston, 21 Jan., F.O. France, of Webster, I, 128. The Protocol was that of 20 Jan., No. 9, cf. above, p. 84.

[25] Granville to Palmerston, 21, 24 Jan., F.O. France; Granville to Grey, 24 Jan., Howick MSS.

Britain would have given up the Prince of Orange's cause, if she were sure that the Belgians really object to him. "But all our accounts represent his party as daily gaining strength, & if they could succeed in establishing him, *tho' by a short struggle*,[26] I must think that it would be the best settlement for all parties." Then follow Grey's pacific assurances, not excluding, however, his pointing out that England's maritime power could destroy France. The best means for continuing cordial relations and security for Louis Philippe's government was the Prince of Orange in Belgium. "For God's sake, therefore, do not reject this hope too peremptorily!" As for the Poles, Grey denounced with indignation the alleged influence of Mme Lieven with him, but did not see any means of intervention.

Should the Orange election prove impossible, any other king is acceptable except one who would only be a French viceroy.[27]

But Granville could only report the current news: the Belgian agent in France, Count de Celles, proposed Marshal Gérard or Lafayette to act as Regent of Belgium. Either solution amounted to a camouflaged reunion; meanwhile the French Government decided to call up the levy of 80.000, which had previously been fixed for April, immediately.[28]

Later reports grew more alarming. On 31st January Sebastiani declared that the French army would enter Belgium if Leuchtenberg were elected; on 3rd February Louis Philippe told Granville that a straight refusal of the Duc de Nemours, if elected, would mean anarchy in Belgium, and therefore must be gradual. At the same time Sebastiani semi-officially informed Bresson, the French member of the Conference's delegation at Brussels, that the King of the French would allow his son to wear the crown of Belgium; and Bresson neither prevented the publication of this letter nor denied its contents. Although immediately after learning of the election of Nemours and – as Palmerston suspected – of the British Cabinet's decision "to require France to refuse acceptance at the risk of war" Sebastiani markedly changed his tone, mutual relations hardly improved. Granville headed the protest of the representatives of the Four Powers against Bresson's declaration to the Belgians that France would not confirm the protocol on the division of the debt of

[26] Italics mine.
[27] Flahaut to Grey, 17 Jan., Grey to Flahaut, 20 Jan., Howick MSS.
[28] Granville to Palmerston, 24 Jan., and above, p. 107.

the Netherlands, and Sebastiani answered that Bresson might have done wrong in making this declaration, but he only expressed the view of the French Government. When Granville complained that this attitude encouraged Belgian opposition to the Conference, Sebastiani reiterated that the Conference did not intend to enforce its decisions; if that was the case, it would unavoidably also bring French troops into Belgium; and it would be difficult to withdraw them, once they had entered.[29]

Mutual opposition grew more distinct as, on the other hand, the French Government continued to accuse the Four Powers, and Britain in particular, of the will to impose the Prince of Orange upon Belgium. The interest displayed by Lord Grey's administration does not seem to have been entirely accidental; that its attitude towards the French Government strengthened this impression is another piece of evidence of strained Anglo-French relations at the time; and when the election crisis in Brussels was over, complaints from Paris became a frequent feature of the Conference; for the rest soon after the January proclamation of the Prince his cause lost a good deal in the eyes of the British Government.

After the Conference's attempt in the Prince of Orange's favour Grey informed Mme Lieven that the Prince had shown him letters from two opposite Belgian parties, both demanding his proclamation to be made in the very same way. But only a few days after the Conference had informed the King of the Netherlands that all the Five Powers backed the canditure of his son, Grey wrote to the Princess that the Orange party seemed to be growing weaker. As the Protocol of 20th January – since repeatedly referred to as the "Bases de séparation" – had been agreed upon, whereby the neutralization of Belgian territory was assured, any candidate the Belgians might be willing to accept would be suitable to the Powers.[30]

Different language was employed between Paris and London. Sebastiani told the British Ambassador that the Prince of Orange would be opposed by force, while Talleyrand, more diplomatically, attempted to get the Conference to exclude the Prince together with

[29] Granville to Palmerston, 31., 4, 11, 12 Febr., F.O. France; Granville to Palmerston, 4 Febr., Palmerston to Granville, 8 Febr., Bulwer, II, 38.

[30] Bagot had audience of William I on 18 January. Grey to Mme Lieven, 16, 22, 30 Jan., *The Lieven–Grey Correspondence*, II, 135, 142, 150. The second letter reads: "We still hear of the Orange party, but it does nothing and does not advance a step."

all members of the Royal families of the Conference Powers from the Belgian election, again arguing that the choice of him would mean civil war; to which Palmerston replied that he doubted "whether civil war would not be the certain consequence of any other choice" but the Prince alone.[31]

In the meantime, Lord Ponsonby increased his efforts, as did Lieven also. Sebastiani complained to the British Ambassador about Krüdener, who continued his action in Belgium. Talleyrand, too, made the presence of this Russian agent the subject of two interviews with the Russian Ambassador.[32]

Krüdener did not consider his mission easy. "Le parti du Prince d'Orange est nombreux", he wrote on 23rd January, "mais comprimé par la canaille."; and two days later, "Le parti du Prince d'Orange est partout, mais n'ose se montrer nulle part." On 2nd February he thought, however, that the revocation of the exclusion of the Nassaus, as far as the Prince was concerned, might be proposed in the Belgian Congress. This would be opposed by France, at least by her agencies in Brussels, in so far as credence could be given by the Belgians to Bresson's declaration that France had the means to support Nemours against the whole of Europe. On the other hand, Krüdener could rely on Ponsonby's support; this envoy expressed to him his belief in the ultimate success of the Prince and promised his co-operation for that purpose. The hesitating Lieven himself was afraid, however, that although the Prince regulated his movements by the advice of the Conference, his father's interference handicapped their work.[33]

It remains to be noted that at this time, between the election of the Duc de Nemours and the final refusal of the offer by his father, a singular solution was put forward by Bresson, and perhaps by others as well, to get rid of the Prince of Orange. Bresson wrote both to Sebastiani and to Talleyrand to express his belief that a proposal made by France to place the Prince of Orange on the Polish throne, even if unsuccessful, "pourra affermir le Duc de Nemours en Belgique". Talleyrand did not consider seriously "une

[31] Granville to Palmerston, 4 Febr., F.O. France; Palmerston to Granville, 1 Febr., Howick MSS.

[32] Granville to Palmerston, 13 Febr., F.O. France; Talleyrand to Sebastiani, 16 Febr., Talleyrand, *Memoirs*, IV, 77.

[33] Krüdener to Lieven, 23, 25 Jan., 2 Febr., Ponsonby to Krüdener, 31 Jan., Lieven to Gourieff, undated but written at the end of January or in the beginning of February. Lie MSS.

pareille absurdité", but remarked that "l'aplomb avec lequel il la faisait valoir doit me faire supposer qu'il se sent appuyé quelque part." This strange combination was indeed brought forward by the young Duc d'Orléans in a conversation with the Polish agents in Paris, and found an echo some months later in Warsaw. Evidence is lacking as to whether it had originated with the Duc, Bresson or someone else.[34]

A different suggestion concerning the connection that France wished to introduce between Polish and Belgian affairs is reported by Bagot. After the election of the Duc de Nemours, de Mareuil, the French Minister in The Hague, repeatedly tried to impress upon the British Ambassador and, as Bagot believed, upon others as well that France, though by no means inclined to accept the Belgian offer, might "so far avail herself of the offer as to make it the ground and instrument of some negotiation as regarded Poland, and endeavour to couple her refusal of the Belgian throne for a French Prince with some effort to secure for the Poles a sort of neutrality similar to that proposed for Belgium." The British Ambassador considered the idea of some importance because of the delay made by the King of the French in his final refusal.[35]

The King of the Netherlands also found in the French failure to introduce their candidate a new hope for the Nassau restoration in Brussels. He developed his views before the British Ambassador. William I thought that the Conference, as they had already largely intervened in the affairs of Belgium and Holland through the protocols concerning territory and finances, might go a step further and "declare that the welfare of Belgium and the peace of Europe forbade a longer continuance of the present state of things and confirming the separation of the Two Countries upon the basis which they had already decreed, to pronounce their federative union with an entirely separate administration of the Belgian provinces, to be permanently delegated to the Prince of Orange, or some other Prince of the House of Nassau in some form, similar to that which existed when they formed a part of the Austrian dominions." To this Sir Charles Bagot replied that this matter depended entirely on

[34] Bresson to Talleyrand, 5 Febr., Talleyrand, *Memoirs,* IV, 53. It should be noted that, as the Polish throne had been declared vacant on 25 January, Bresson could hardly have learnt of it, when he made his proposition. The Duc d'Orléans repeated it on the 8th, T. Jungfer, *Die Beziehungen der Julimonarchie zum Königreich Polen,* Berlin, 1909, p. 23.

[35] Bagot to Palmerston, 11 Febr., Bagot MSS.

the attitude of France – if she wished for peace, the King's suggestion could be brought forward and he certainly favoured it. In his report, however, Bagot regretted that William I insisted on retaining the sovereignty of Belgium in his own person.[36]

Palmerston, however, did not give any consideration to this proposition. Earlier, and even before he learnt of the election of Nemours, he had been afraid that the Orange party showed no courage and that "the Prince's chances are nearly gone"; to Lieven he wrote hopefully but expressed his opinion that the Prince of Orange must choose between Holland and Belgium, or he would lose both countries – a suggestion largely opposed to William I's project, though not absolutely so. The Foreign Secretary took care to point out that Ponsonby's efforts in favour of the Prince must not be known. Again, a few days later, and at the very moment when the election of Nemours became known, Palmerston observed that the Prince's election would, contrary to the policy of France, tend to keep the two countries together. "However", he added, "the Orange cause seems to become hopeless." [37]

All this time the two envoys who had arrived from the City of Ghent to plead with the Conference for an Orange restoration were staying in London. They noticed the special interest of Palmerston and Lieven for their mission. It was the Russian Ambassador to whom they addressed an *exposé* on the possibility of the Brussels Congress being persuaded to elect the Prince of Orange to the throne of Belgium. Lieven's agent in Belgium reported at this time (on the 15th): "Je pousse aux pétitions [to the Congress] en faveur du Prince." [38]

Also at the same time, however, Talleyrand succeeded in obtaining from the British Foreign Secretary the promise that Ponsonby would no longer interfere in favour of the Prince of Orange; Ponsonby had of course no instructions which would warrant his open intervention for this purpose, but his sympathies for the Prince had for some time been undisguised, and caused much anxiety to the French Government. Palmerston's promise was, therefore, much

[36] Bagot to Palmerston, 13 Febr., Bagot MSS.

[37] Palmerston to Bagot, 2 Febr., Bagot MSS; Palmerston to Lieven, 3 Febr.; "Mille remerciments [for documents on the Orangist movement]. Tout paraît aller bien et je souhaite bien que tout cela réussit à notre gré." Lie MSS; Palmerston to Grey, 6 Febr., Howick MSS.

[38] The Ghent envoys to Lieven, 15 Febr., Krüdener to Lieven, 11 Febr., Lie MSS. cf. above, p. 83.

appreciated. Yet, when a few days later Ponsonby was the subject of another note from Sebastiani to the Ambassador in London, the latter reminded the Minister that though Palmerston did not believe in the Prince's success any longer, he always maintained that the Prince would be the best solution for Belgium. In any case, even Talleyrand considered his candidature as still being held under serious consideration.[39]

The Netherlands Ambassador went even further, believing that serious consideration was also being given to the plan of William I, whereby, as stated, the sovereignty of Belgium was to remain with the King of the Netherlands. Falck's analysis of the situation with regard to the Belgian crown ran as follows: all the four Powers would welcome the solution on the lines proposed by William I, but neither Britain nor the other three Powers were prepared to uphold it against French opposition, which had been growing lately in spite of Louis Philippe's efforts. Besides, the Eastern Powers probably wanted to play for time until the situation in Italy and Poland should clear.[40]

This was not quite true with regard to Vienna where even before the Nemours crisis the candidature of the Prince of Orange had been renounced, which supplied the French Government with a new argument against the Prince. But the success of the firmness of Britain in opposing the Duc de Nemours filled the other two Courts with fresh hopes for the Prince's success and for the idea of the Belgians remaining under their lawful ruler. And Palmerston, if not prepared to uphold the Prince with any special efforts, was nevertheless equally convinced that by opposing the Prince France intended something else than merely studying the wishes of the Belgians. "France seems daily to be drawing aside the mask", he wrote to Grey, "in proportion as her military preparations advance." [41]

[39] Talleyrand to Sebastiani, 19, 16 Febr., Talleyrand, *Memoirs,* IV, 67, 77-8.

[40] Falck to Verstolk, 25 Febr., Falck Papers, R.A.

[41] Maison to Sebastiani, 4 Jan., A.é. Autriche; Palmerston to Lieven, 18 Jan.: "...on a découvert à Paris que l'Empereur d'Autriche proteste contre tout *Prince Protestant,* ce qu'on ne savait pas à Vienne le 7me de ce mois, la date des dernières dépêches de Cowley." Lie MSS. This dispatch was almost exclusively devoted to Polish affairs, cf. above, p. 94; for Mortier's and Mortemart's reports see above pp. 104 and 105. On the occasion of Sebastiani's statement in the Chamber that the Belgian fortresses must not be allowed to fall into the hands of the enemies of France, Palmerston commented to Grey: "Does he mean the Prince of Orange?" Then immediately follows the quotation in the text of this page. Palmerston to Grey, 15 Febr., Howick MSS; cf. also the widely quoted

But in the beginning of March Palmerston did not hesitate to tell even the Russian Ambassador that the Prince of Orange was daily losing ground in Belgium. By this time also St. Petersburg was about to learn from Sebastiani's dispatch that Ponsonby "a reçu de son Gouvernement l'ordre de ne plus s'occuper du Prince d'Orange." Lord Ponsonby himself, however, still believed that the Belgian army was entirely for the Prince. He was surprised to see the Belgian Government aiming at war with Holland. But he did not underestimate the support the Belgians were likely to obtain from France – indeed his report that Belgian border fortresses were fully provisioned, but not garrisoned, had caught Palmerston's attention a few days earlier. He eventually summed up his three months' experience at Brussels as follows: "I strenuously recommend to Palmerston to give up all idea of acting here for the present and to direct his whole attention to the working of Belgian affairs at Paris, where alone originate the difficulties in the way of the settlement of this country. Take away the fears of France, and I would not ask many days to carry any measures our Government might desire to have carried." [42] At that time, at least, Palmerston's own experiences may also have brought him to some such conclusion.

After the Nemours crisis was over the Belgians, aware of the difficulties of a Royal election, postponed the election of a King, and on 25th February the Congress chose Baron Erasme Surlet de Chokier to act as Regent of the country.

The French Government hastened to accredit General Belliard to him – since Bresson, after all that had passed at the time of the election of Nemours could no longer continue in Brussels. This appointment caused some surprise. The mission of this officer was not "simply diplomatick", thought Granville: no doubt, his instructions "are prepared with a view to the possibility of the Low Countries being the theatre of military operations." [43] Other Powers were not likely to follow France's example; the Provisional Govern-

---

letter to Granville, of the same date, to make Sebastiani understand "that our desire for peace will never lead us to submit to affront either in language or act." Bulwer, II, 41.

[42] Palmerston to Lieven, 2 Mar.,: "Je crains que le Prince d'Orange ne devienne de jour au jour plus éloigné du trône belge." Lie MSS; Sebastiani to Mortemart, 20 Febr., A.é. Russie; Ponsonby to Bagot, 2, 6 Mar. (the quotation is taken from the second letter), Bagot MSS; Palmerston to Granville, 23 Febr., Bulwer, II, 45.

[43] Granville to Palmerston, 4 Mar., F.O. France.

ment at Brussels attempted, however, to enter into diplomatic relations with various Governments, and special envoys were to announce Surlet's appointment; Van de Weyer, who had already been in London as a semi-official agent of the Brussels authorities, returned to the seat of the Conference. The Regency of Belgium, however, did not dissipate the fears of the Powers. "I suppose from him who is sent to England", commented Granville on the last measure of the Provisional Government of Belgium," you will learn in whose name Choquier de Surlet [sic] is regent, and whether his appointment supersedes the election of the Duc de Nemours." Nevertheless, the French Government would not admit that the belated refusal of the Belgian offer by Louis Philippe had not been meant to be final; the British Ambassador reported in the same letter that Sebastiani "continues to assure me in a whispering confidential tone that he is quietly working for Leopold of Coburgh." [44]

This was not enough for the British Government, and Palmerston did not wait for Van de Weyer's arrival. He requested the Belgian Government to state officially whether Surlet's nomination meant the cancellation of the election of Nemours, and in his answer to the Paris Ambassador he showed his annoyance. "I wish the French Government would make up their mind", he wrote "to act with good faith about Belgium, and we should settle the matter in three weeks; but the men in power cannot make up their minds to be honest with stoutness or to play the rogue with boldness. Might they not be reminded that when the Russians have conquered Poland, which (were it not for the ill-concealed spirit of aggrandizement of France) I should say I am afraid they will, the tone of Russia about Belgium will be different from what it has been, and that Prussia and Austria will probably be swayed by her influence? If, therefore, the French ministers mean what they say about Leopold and are not amusing us, they had better not delay the matter till the Poles are at the foot of Nicholas." [45]

Evidence is lacking whether this meant that the British Government, or Palmerston personally, were otherwise prepared, at that moment, to enter in some rôle in the contest on the Vistula. Though secretly, Palmerston still did not refuse to see the Polish envoy,

[44] Granville to Palmerston, 7 Mar., Palm MSS.
[45] Palmerston to Ponsonby, 9 Mar., F.O. Belgium; Palmerston to Granville, 9 Mar., Bulwer, II, 48.

Wielopolski, whose arrival in January had received considerable publicity. Only a few days before Palmerston wrote the above, he had told the Polish envoy that though generally speaking the Cabinet was well-disposed towards Poland, they needed the support of Russia against France; he also added that the friends of Poland in France were also the most warlike, and in this way increased the anxiety of other Powers.[46]

III

In the beginning of February Diebitsch's main army crossed the frontier between the Kingdom of Poland and Russia, and advanced unopposed upon Warsaw.[47] Eventually came the long awaited event: on 25th February the battle of Grochów, in the neighbourhood of the left-bank suburb of Warsaw, Praga, took place; its outcome had been predicted long beforehand, and all the circumstances could not but confirm this. The Prussian Consul in Warsaw immediately informed Berlin, so that on 3rd March Chad, Mortier and Trautmannsdorff, the Austrian Ambassador, sent couriers to their respective courts, informing them of the disaster to the Polish army; similar accounts left Vienna and St. Petersburg a few days later. [48]

The correctness of the news was not questioned in Paris nor in London. The people of Paris reacted with an assault on the residence of the Russian Ambassador; the French Government had to take measures of its own as well. Sebastiani wrote immediately to Berlin, Vienna and St. Petersburg. The Prussian and Austrian Courts were requested to intercede for the defeated; the French representatives were instructed to insist on the maintenance of the stipulations of the Treaty of Vienna concerning Poland. And Mortemart, whose task was the most difficult, was reminded of the "question d'une haute importance, quel sort le gouvernement russe réserve à la Pologne." The Minister admitted the delicacy of the subject, but insisted that

[46] Wielopolski to Plater, 19 Mar., reporting his interview with Palmerston on the 5th. B.P.

[47] The battle of Stoczek, referred to above (p. 105), had been fought between small Polish and Russian corps.

[48] Mortier to Sebastiani, 3, 5 Mar., A.é. Prusse; Mortemart to Sebastiani, 5 Mar.,: "On attend des détails ultérieurs sur cette affaire qui porte un coup terrible aux moyens de résistance des Polonais et sera sans doute suivie bientôt de l'action décisive qui doit terminer cette courte mais désastreuse lutte." A.é. Russie.

it must be mentioned. He also expressed to Granville his hope that Mortemart might be supported by Lord Heytesbury.[49]

The ground was otherwise hardly prepared for co-operation. The news of the Polish disaster was on its way and arrived in London on the evening of the day that Palmerston, annoyed at French duplicity, predicted its inevitable consequences on the fate of Poland and on the attitude of the Eastern Courts. The news could not fail to influence London. That same day the City gave a dinner in honour of the Polish envoy, Marquis Wielopolski. Princess Lieven, a few days later, described it as a sort of failure: "Burdett qui devait y présider, n'y est pas venu. Il n'y a un de nous connus [sic] que Mr. Hobhouse. Lord Grey m'a dit hier", continued the wife of the Russian Ambassador, "qu'il lui avait destiné et promis une place dans le Gouvernement, mais que ce dîner rend impossible de la lui donner." Hobhouse himself recorded the occasion and its consequences.[50]

The Lievens received "bonnes nouvelles" from Alopeus, the Ambassador in Berlin, during a dinner at Lord Londonderry's. All the Tories congratulated them, also Lord Melbourne, who was present. Elsewhere slightly different comments were uttered. Palmerston on learning of the event is reported to have said that he had no doubts about the moderation of Nicholas, and also, that governments had to suppress their sympathy for the Polish cause under these circumstances when every throne and authority had been menaced by revolution, and when the support of Russia was required to check revolutionary France. Grey's alleged words to Talleyrand, of which Princess Lieven rendered an account to the Russian capital, were: "All is ended – and it is most unfortunate." [51]

[49] Sebastiani to Mortemart, Maison and Mortier, 8, 9, 10 Mar. respectively, A.é. Russie, Autriche, Prusse; Granville to Palmerston, 11 Mar., F.O. France.

[50] Cf. above, p. 92; Mme Lieven to Alex. Benckendorff, 13 Mar., Lie MSS; Hobhouse, on the occasion of the offer of the Secretaryship for War, 31 March, 1832, noted that he would have had it earlier "had it not been for my speech at the Polish dinner." Hobhouse, *Some Account of a Long Life,* II, 167.

[51] Mme Lieven to Alex. Benckendorff, 13 Mar., Lie MSS; Falck and Van Zuylen to Verstolk, 10 Mar., R.A.; *The Lieven–Grey Correspondence,* II, 184, Mme Lieven to Grey, 11, 12 Mar; Grey to Mme Lieven: "There is not one word of truth in the exclamation which is said to have escaped me. I have never concealed from you, however, that I feel a good deal of compassion for these poor Poles. But it has never influenced the conduct which my public duty prescribed to me." *The Lieven–Grey Correspondence,* II, 185; the same text, translated into French, is printed in Nesselrode, *Memoirs,* VII, 177.

In the meantime, the completion of the Polish disaster still remained to be recorded. Very shortly after the first news of the battle, the Western capitals were to hear reports of a different kind. The first came from Berlin. Mortier was surprised to find that Praga (the left-bank suburb of Warsaw) had not yet been taken, and that the Polish army was ready to repel a new attack. A few days later he reported that Marshal Gneisenau had left for Poznan to take the command of the army destined to assist, if necessary, in the disarming of the Polish forces; but to the French *Chargé d'affaires* the news from Warsaw did not appear to warrant this step. Moreover the Russian commander, Diebitsch, had sent back some of the Polish prisoners, "auxquels il a donné deux ducats chacun, en leur remettant des proclamations, où il ne traite plus les Polonais de rebelles, mais parle au contraire de leur belle et valeureuse nation." Chad, after some confused accounts, had heard that Polish losses amounted to 30.000 men, but remarked that the information came from a source unfriendly to the Poles.[52]

Accounts from St. Petersburg seemed less equivocal. Heytesbury, after his note of the 5th on the Russian victory, believed on 8th March that there was disorder and hopelessness in Warsaw. "Les nouvelles de France, malgré leur importance" – at that time the Nemours crisis was becoming known in St. Petersburg – wrote the French Ambassador, "semblent passer ici sans être aperçues de l'Empereur, tant il est préoccupé par celles de la Pologne." Of the ultimate result Mortemart had no doubts. He now reckoned Polish resistance would last another month, and foresaw heavy losses for the Russians. Nicholas would be glad to have the affair ended with a reconciliation with the Poles, but would not himself make the first step. The French Ambassador was anxious to avail himself of every opportunity to bring about a peaceful solution, "mais l'acharnement des deux partis m'en présentent bien peu." A week later he had news of new levies of recruits: this measure, he thought, would empty villages and drain the financial resources of both the state and individuals. "Je considère la Russie comme paralysée dans ce moment pour les affaires de l'Europe, s'il fallait agir.", concluded the Ambassador.[53]

[52] Mortier to Sebastiani, 5, 10 Mar., A.é. Prusse; Chad to Palmerston, 6 Mar., F.O. Prussia.

[53] Heytesbury to Palmerston, 5, 8 Mar., F.O. Russia; Mortemart to Sebastiani, 9, 16 Mar., A.é. Russie.

This conclusion was amply supported from Vienna by Cowley's report at a later time, when more information became available. "There are grounds to believe", wrote the British Ambassador, "that Marshal Diebitsch has no reason to boast of his success on the 25th ultimo. On that day the left wing of the Russians (as I am assured upon the best authority) was defeated, and on the right the Poles had advantage until two o'clock in the day when General Chlopicki, being severely wounded, was compelled to quit the field." This caused disorder, and eventually, "a charge of the Polish Uhlans facilitated the retreat of the army into Warsaw, the tête du pont being, however, sufficiently guarded." But the Ambassador believed, in spite of his appreciation of the fine qualities of the Polish army, in the ultimate success of the Russians, "both on account of their superior numbers and the moral support, afforded them by the Austrians and Prussians, which entirely prevents the Poles from deriving any succours from abroad." On the other hand, the prolongation of the rising was absorbing enough to prevent Nicholas from assisting any of his allies in the event of a war with France. This reasoning was supported by an account of Russian losses.[54]

This authoritative information could not have arrived in London before the end of March. Even before that, however, leading opinions in England seemed to be almost as cautious. Grey observed, within a week of the first news of the great battle, that in spite of victory in the field Diebitsch no longer found it possible to try to carry Warsaw from behind the Vistula and had to embark upon a long course of operations, involving the crossing of the river at a considerable distance from the capital of the Kingdom. Wellington admitted some influence of Diebitsch's operations upon the state of affairs in Europe, but he thought that the international situation urgently demanded his further action, which the Russian commander indeed seemed to have been preparing in spite of the adverse season. "Je crains que son armée n'en souffrira terriblement", commented the Duke to Princess Lieven, who sent his letter to St. Petersburg. Yet, she did not wholly perceive the warning; she wrote home that the Polish affairs did not occupy the public, "tant il est vrai qu'ici, du moins, le succès efface tout, et que l'homme battu est pour les Anglais un homme mort. D'ailleurs,

[54] Cowley to Palmerston, 19 Mar., F.O. Austria.

leurs propres affaires les absorbent tout entier." It rested with the Russian reader to judge whether the last sentence conformed with the preceding.[55]

If in London, occupied with the Reform Bill, comparative indifference with regard to Poland prevailed, this could not be the case in France. The French Government was already alarmed at the development in Italy; Louis Philippe's and Sebastiani's trying to abstain from a more energetic policy in that quarter was the final blow to the existence of the Cabinet of Laffitte,[56] while general opinion maintained that "as soon as Russian has finished Poland, France will be attacked" – as the influential Rothschild put it. The position of the Government was undermined in the country by their passive attitude in Polish affairs – thought the British Ambassador – they had to do something in Italy to restore their prestige. It was not enough that Government circles as well as the liberal Sardinian Court were convinced that, whatever Nicholas' intentions, he had become unable to interfere in affairs beyond his own empire, which again would make Austria more pacific.[57]

The Polish revolution was therefore during March three times the subject of discussions in the French Chambers. The Warsaw papers of the Grand Duke Constantine were quoted as evidence of the Russian plans to attack France; but the Government maintained that this evidence was not sufficient; though it must be remarked that the French Consul in Warsaw, whose dispatches were one of the channels through which Constantine's papers penetrated into Paris, commented on the "Polish" debates that there seemed to have been no project of attacking France, but that Nicholas had certainly been going to intervene in Belgium, which of course would have forced France into war. In London as well as in St. Petersburg a curious story was told that during the speech of Lafayette Sebastiani 'tenait dans sa poche' evidence of Lafayette's having been the organizer of the Polish revolution; a wink from the speaker's podium prevented the Minister from using it, as Lafayette

[55] Grey to Mme Lieven, 15 Mar., *The Lieven–Grey Correspondence,* II, 188; Wellington to Mme Lieven, 12 Mar., Wellington, *Supplementary Despatches,* VII, 41; Mme Lieven to Nesselrode, 18 Mar., Lie MSS.

[56] Cf. *Mémoires de Laffitte,* ed. P. Duchon, 1936, pp. 267 sq.

[57] James Rothschild to Salomon Rothschild, 13 Mar., E. Corti, *The Reign of the House of Rothschild,* 1928, p. 19; Granville to Palmerston, 18 Mar., F.O. France; Barante to Sebastiani, 12 Mar., Barante, *Memoirs,* IV, 130.

seemed then to propose to reveal that his action was preconcerted with Sebastiani.[58]

The parliamentary difficulties were increased by a serious development in Belgium with regard to the Luxemburg question. It had been noted that the population of the Grand Duchy joined hands with Brussels, and that the Conference was unable to find a solution which would make the union of Luxemburg with Belgium possible. But the Provisional Government seized the first occasion to reaffirm the general will to achieve this union: one of the first great acts of the Regent was the proclamation to the people of Luxemburg saying that in spite of the decision of the Conference (the "Bases de séparation") and of the proclamation of the grandducal Governor, they would belong to the new Belgian State.

This was a challenge not only to the Conference, but to the German Confederation as well. In reply, the German Diet agreed immediately to send Federal contingents to occupy the Grand Duchy, a step already under consideration for almost half a year. For his part, Palmerston refused to accept the credentials of the Belgian envoy, Van de Weyer, and informed him that the Conference would no longer be able to prevent the Federal occupation. The French Government also disapproved of Surlet's proclamation: if the Federal measures were put into force, hostilities in the Grand Duchy were unavoidable, and that again could not be tolerated by France. Belliard, however, was instructed to make the Belgians understand that France, too, would not prevent the Federal occupation.[59]

With the precarious situation in Italy and in Poland, with the nightmare of an Orange restoration in Belgium, the French Government necessarily remained on the defensive in the Luxemburg question; but even the new Cabinet of the moderate Casimir Périer could not give up the essentials of French politics since the July revolution. Sebastiani could only succeed in "amusing" the British Ambassador with the subject of Prince Leopold's candidature for the throne of Belgium: both Palmerston and Granville were scep-

[58] Dutkiewicz, pp. 88-9; Durand to Sebastiani, 5 Apr., A.é. Pologne; Mme Lieven to Alex. Benckendorff, 31 Mar., describes an account from Grey of the alleged incident in the Chamber with Lafayette's correspondence. Sebastiani "allait la produire lorsqu'un signe menaçant de Lafayette l'y a fait renoncer. Ce signe était pour lui rappeler que c'était en sortant de chez lui [Sebastiani] que Lafayette était allé expédier ses ordres dans le tems au comité révolutionnaire de Varsovie." Lie MSS.

[59] Franqué, p. 72 and 79; Sebastiani to Belliard, 15 Mar., Belliard, *Memoirs*, II, 21.

tical about French intentions in this regard, in so far as they considered Sebastiani's declarations as a French defence against the Prince of Orange, and accordingly Granville did not propose to show his support for Leopold. There was hope of a change as Périer was installed on 13th March. The British Ambassador was glad to find that the new premier's disposition was peaceful. However, he still saw himself compelled to record Périer's "obscure hint" that a little acquisition of territory could greatly strengthen the popularity of the new Cabinet and the whole peace party: would not a part of Bouillon – which had not originally belonged to Luxemburg – lend itself well for that purpose? [60]

Neither did Sebastiani satisfy the Ambassador by telling him that Belliard and Bresson had dined with the Regent on the eve of his proclamation to the Luxemburgers, and yet were told nothing about the proposed move. The Minister, on the other hand, was able to recount with glee the intercourse between the Regent and Lord Ponsonby: the British delegate was said to have complained of French influence with the Provinsional Government, to which Surlet de Chokier replied "that the influence of England being entirely employed to overturn the present Government with a view of placing the Prince of Orange on the Throne, it could not be expected that he [the Regent] could feel any reliance on the friendship of the English Government." However, to the British Ambassador Sebastiani dared not yet support Belliard's declaration that the Prince of Orange would be opposed by France by force – a declaration which Granville was reported to have called an act of interference – but he took the opportunity to mention that Ponsonby "was concerting with the persons in opposition to the Government of the Country" as if echoing the much more positive statement of the Belgian Regent.[61]

A different tone could be used with regard to Brussels. Later in March, being no doubt in possession of the supplementary reports from Poland, Sebastiani warned the Belgians to be on their guard as far as the Dutch were concerned, but not to be afraid of a general war: "La Prusse ne veut pas la guerre. La Russie est occupée en Pologne, l'Autriche en Italie." France, would, therefore, oppose an Orange

[60] Granville to Palmerston, 14, 21 Mar., Palm MSS; cf. another reference to Bouillon, above, p. 73 and below, p. 144.; See Appendix VIII.

[61] Granville to Palmerston, 18 Mar., F.O. France; Fagel to Verstolk, 19 Mar., R.A.

restoration by force, but the Belgians must not start hostilities themselves, not intimidate the Conference – as they already had done by the proclamation.[62]

In accordance with his view of the Eastern Powers thus expounded Sebastiani tried to influence the most interested and seemingly the weakest partner of the Three, that is to say Prussia. The French representative had already been instructed to explain French policy with regard to the new complications in Belgium. Surlet's proclamation was disavowed by France, but warlike projects were also entertained by the King of the Netherlands. France would not allow any military operations in the Low Countries. Firstly, no intervention would be tolerated in Belgium, not even one by the Prince of Orange, "sous quel titre qu'elle pût avoir lieu." Secondly – "la question du Luxembourg n'est pas moins délicate:" France recognized the rights of William I in the Grand Duchy – with the notable exception of Bouillon – but hostilities on the part of the Dutch and Federal troops "sur un territoire aussi rapproché de nos frontières compromettraient véritablement le repos de l'Europe." Sebastiani's overture was directed to the right quarter: even before this dispatch reached Berlin, Ancillon had told the French *Chargé d'affaires* that Bouillon would, for the time being, be exempted from military measures.[63]

This was, however, only a slight concession and other Powers did not conceal from the French, that in their view the intensivity of the resistance offered by Brussels depended on the attitude of France. Metternich could permit himself to suggest that William I should be allowed to reconquer the province, but he too made an exception for Bouillon. Bernstorff, more cautiously but quite plainly, told the French *Chargé d'affaires* that if France advised the Belgians to submit, he was sure that they would do so. Berlin felt the support of London in this question, and Mortier thought that a military occupation of Luxemburg could no longer be delayed.[64]

The French Government was not only unable to allow military operations to take place in Belgium; the pressure of public opinion made them equally reluctant to influence the Belgians in the way the other Conference Powers thought necessary. Though they recog-

[62] Sebastiani to Belliard, 24, 26 Mar., Belliard, *Memoirs*, II, 24-27.

[63] Sebastiani to Mortier, 22 Mar., Mortier to Sebastiani, 23 Mar., A.é. Prusse.

[64] Maison to Sebastiani, 21 Mar., Mortier to Sebastiani, 30 Mar., A.é. Autriche, Prusse.

nized the rights of William I to the Grand Duchy, they professed themselves unable to make a public declaration to this effect. Diplomatic interviews, therefore, continued to by-pass the real problems; Louis Philippe pointed out to Granville the practical difficulties in restoring Luxemburg to William I as it had no common frontier with Holland. The King seized this opportunity to mention to the Ambassador, that prior to 1815 Luxemburg and Belgium were always (sic) united.[65]

The pressure of the other Powers grew and created an unavoidable dilemma for the French Government. Much as they realized that the Powers were not very eager to force the Belgians into accepting the "Bases de séparation", it was equally clear that any hostilities commenced by the Belgians would bring about disastrous results, not in the least to France herself. There seemed to be no possibility of a compromise, and the "Bases" had to be presented at Brussels as "ce qui importe à la Belgique... son indépendance... sa séparation de la Hollande", as important for France as they were for Belgium, and for that reason Sebastiani tried to argue with the British Ambassador against any compulsion with regard to the Belgians, however stubborn they might be. But Granville merely observed that even the independence of Belgium was as yet provisional.[66]

This notion, however, was only too well grounded in Paris and, in one way or the other might provide France with a solution. The idea of partition had already been put forward before; Sebastiani, too, did not wait for this comment to use the argument which perhaps could assure the acceptance of the "Bases" in Brussels; in one of the interviews with the aide-de-camp of Belliard intended to promote this object the Minister remarked incidentally: "les Belges n'ont que des idées folles; qu'ils y prennent garde, on les partagera." The diplomats in Paris had reason to believe that this was not just an incidental threat, and that the King was not hostile to the idea. All that Sebastiani told Granville, however, was that the best means of preventing hostilities on either side would be a military occupation of Belgium by the French and Prussian forces. This of course was unlikely to be attractive to the British Government.[67]

[65] Granville to Palmerston, 1, 4 April, F.O. France.

[66] Sebastiani to Belliard, 4 Apr., Belliard, *Memoirs,* II, 36; Granville to Palmerston, 4 Apr., F.O. France.

[67] Répecaud to Belliard, 1, 2 Apr., Belliard, *Memoirs,* III, 22-26; Werther to

In the meantime the problem of the Belgian crown came no nearer solution. The French Government did not press this matter; but it was no longer prepared to compromise on the candidature of the Prince of Orange, not even towards Britain. All this time the presence of Ponsonby in Brussels was for Paris a source of constant anxiety and frequent complaints. On the very day that Ponsonby came to the conclusion that an Orange restoration in Belgium was no longer possible, Sebastiani denounced him as being "undefatigable" in his efforts to promote this cause, thus bringing about bad relations with Belliard. This accusation was flung at the British Ambassador in one of Sebastiani's bad tempers: if civil war should break out in Belgium as the result of this situation, he declared, the French army would enter the country. But Granville took as serious a view. "I cannot but express my fears," he told the French Minister, "that you would in that case find yourself at war with all the great Powers of Europe." [68]

IV

The new Polish agent in London, Count Walewski, chose this moment to approach Palmerston again. The substance of his statement to the Pole Palmerston mentioned in a few sentences to Granville; Walewski made an extensive report. The Foreign Secretary explained to him that Britain would only try to maintain the basic arrangements of the Treaty of Vienna and could not be expected to help to make Poland entirely independent, weighty reasons preventing the British Government from a breach with St. Petersburg: "Aujourd'hui, ce n'est pas la Russie que nous craignons", said Palmerston, "c'est la France; nous avons même besoin que la Russie... soit forte et puissante pour l'opposer à la France. Je n'ai aucune confiance dans le cabinet Français. J'ai eu bien de me convaincre de toute sa duplicité dans les affaires de la Belgique, il nous a appris à la connaître. Il ne respire qu'agrandissement et esprit de conquête, et dans ce moment-ci nous ne pouvons avoir autre chose à la vue que de le contenir." Walewski replied that France was now unable to make war, but Palmerston maintained

Frederick William III, 6 Apr., Franqué, p. 88; Granville to Palmerston, 8 Apr., F.O. France.

[68] Ponsonby to Bagot, 28 Mar., Bagot MSS; Granville to Palmerston, 28 Mar., F.O. France; the Prussian Minister in London believed that Britain "officially" supported the Prince of Orange till the end of March. Bülow to Frederick William III, 29 Mar., 2 Apr., Franqué, pp. 83-4.

that if unpreparedness is the only reason for a government for keeping the peace, it should be treated in the same way as if it had declared war. He further said that Périer was more trustworthy than others in the French Cabinet, but he, Palmerston, was most suspicious of the Palais Royal and of people like Soult and Sebastiani. "La France", continued the Foreign Secretary, "est l'obstacle qui avait empêché particulièrement de nous mêler de vos affaires." Nor was Palmerston inclined to share Walewski's opinion that Austria would be the first Power to be attacked; if there was to be war, he thought, hostilities would begin on the Belgian border, and Prussia, against whom the Poles had far greater grievances than against Austria, was therefore to be supported by Britain. Palmerston closed the audience by promising further discussion on the subject at a later occasion. The Polish envoy then went to Talleyrand and confided parts of this conversation to him. The French Ambassador belittled Anglo-French difficulties, but left Walewski with a poor impression as to his sincerity.[69]

In the beginning of April, longer than the Palais Royal, the Cabinet at Westminster still remained under the influence of the battle of Grochów, though both Cowley's account and that of Walewski confirmed that the Polish defeat had not been as final as at first represented. Grey considered the Polish cause hopeless, as did Palmerston, although the latter wondered at the losses sustained by Russia. Grey expected prolonged resistance and greatly regretted the inevitable bloodshed. He somehow coupled these opinions with his conviction that a fair deal for Poland would check Russia's aggrandizement and bring reassurance to Europe on that side.[70]

Though, as has already been seen, Grey continually made

[69] Palmerston to Granville, 29 Mar., Bulwer, II, 61; Walewski to Plater, 29 Mar., B.P.

[70] Hobhouse, *Recollections of a Long Life,* IV, 99. Hobhouse saw Grey on April 3. Grey "could not receive Wielopolski in his public capacity, he should be glad to see him as a private gentleman, just as he did Count Walewski, in consequence of a letter he had received from Czartoryski." (Meanwhile Wielopolski had left England; Walewski's "public capacity" was concealed); Grey to Sir Herbert Taylor, 5 Apr.: "Count * has been here for some days . . . his account [of the battle of Grochów] corresponds exactly with that which you may remember to have read in one of Lord Cowley's dispatches [of 19 March, cf. above, p. 121]", same to same, 6 Apr., of the necessity of a fair deal for Poland, *The Correspondence of the late Earl Grey with H.M. King William IV and with Sir Herbert Taylor,* 1867, pp. 208 sq. The name of Walewski is replaced in the text by an asterisk as at the time of publication he was still living after a distinguished public carreer in France under Napoleon III.

similar hints in his correspondence with Princess Lieven, not much was done officially to bring about such a solution. Palmerston's instructions to Heytesbury, written late in March, though sent with a view to a speedy termination of the contest by a Russian victory, laid stress upon some points not yet mentioned by France. He maintained that a distinction must be made between a rebellion and a civil war, the latter being the case in the Kingdom of Poland; while it should likewise be remembered that the Treaty of Vienna also guaranteed certain national institutions to the Polish provinces incorporated in Russia. Palmerston hoped that Russian advantages in Poland would bring about uneasiness in the two other Eastern Powers. Lastly, though Polish affairs required cautious and confidential treatment, Heytesbury was allowed "to risk even a little offence", if any good for the Poles would seem likely to be achieved in this way. The Ambassador was also instructed to concert his steps with the Duc de Mortemart.[71]

The substance of these instructions was soon known both to the French Government, which was especially expectant of British support in everything that concerned Poland, and to the Russian Embassy in London. It was the second time that Mme Lieven had regretfully to record the official interest of London in Poland, while unofficially, it had been a feature of her letters from Grey from time to time, and at a time when things in Poland looked bad for the revolted.[72]

In the Russian capital some discomfort was felt both about politics abroad and about the state of things in Poland, since it had become clear that the battle of Grochów had not resulted in the expected outcome. In the second fortnight of March there was a kind of suspension of hostilities on the Vistula on account of bad weather; this situation was very much deplored by Nicholas. Neither the developments in Italy nor the slow movements of the London Conference enjoyed his approval, but those questions hardly inspired

[71] Palmerston to Heytesbury, 21 Mar., F.O. Russia, 22 Mar. (priv.), Heytesbury Papers.

[72] Sebastiani to Mortemart, 31 Mar.: "Le sort futur de la Pologne n'inspire pas en Angleterre un moins vif intérêt qu'en France." A.é. Russie; Duchesse de Dino to Barante, 31 Mar.: "La lutte, en se prolongeant, donnera, j'espère, aux négociations le temps de produire leur effet. M. de Mortemart trouvera, à cet égard, appui dans son collègue britannique." Barante, Memoirs, IV, 168; Mme Lieven to Alex. Benckendorff, 31 Mar. "C'est toujours de Heytesbury que je vous conjure de prendre garde; ne lui laissez jamais soupçonner, ni à personne, que je sais ici ce que je ne dois pas savoir". Lie MSS; Grey to Mme Lieven, 29 Mar., *The Lieven-Grey Correspondence,* II, 199.

any interest because of Poland; the growing hatred of the Russians towards the Poles prevented any compromise, as did, according to the French Ambassador, the instigations of Austria and Prussia, while moreover the former could not be relied upon because of the compromising information contained in the Warsaw papers of the Grand Duke Constantine. The Austrian Ambassador reported that the Russian Government had lost all hope of recovering those papers, and was therefore reluctant to make the first step towards improving mutual relations.[73]

Reports from Vienna made this reluctance still clearer to any observer; Mortemart could not have known of Metternich's attitude in Polish affairs, when the Chancellor was approached by the French Ambassador in Vienna with regard to the preservation of the Treaty of Vienna stipulations after the Polish defeat. Metternich "s'est exprimé sans détour" wrote Maison, "sur la combinaison politique qui n'a plus fait de l'ancien Royaume de Pologne que des provinces Prussiennes, Autrichiennes et Russes." The Chancellor also seized this opportunity to inform the French Ambassador of documents he had in his archives proving that the Empress Maria Theresia had been very much opposed to the partition of Poland. Maison attached some importance to this statement, though the objects of his *démarche* had not been achieved, neither with regard to Poland, nor to disarmament, for which Metternich had a project ready of a congress at Prague, but only "dès que le dénouement de la lutte polonaise le permettrait."[74] Austria did not seem likely to commit herself firmly to one side or to the other.

More interesting for Nicholas were the early Polish attempts to sound his disposition towards a reconciliation. Those made in Poland received no publicity abroad. The request to the King of Prussia, in the second half of March, could not remain entirely secret. The new Commander-in-Chief, General Skrzynecki, let it be known that the Polish army would submit on condition that the constitution of the Kingdom was maintained. Mortier must have heard some rumours of this attempt; he believed, moreover, that negotiations were already taking place and reported some details: con-

[73] Mortemart to Sebastiani, 26, 29 Mar., A.é. Russie; Nicholas to Constantine, 19, 28 Mar., Correspondance, p. 139, 144; Ficquelmont to Metternich, 22 Mar., Sz. Askenazy, "Polskie zabiegi dyplomatyczne w czasie Powstania Listopadowego" (Polish Diplomatic Endeavours during the November Rising), in *Biblioteka Warszawska*, 1902, III, 444.

[74] Maison to Sebastiani, 21 Mar., A.é. Autriche.

stitution and amnesty were agreed upon, only the army presented a major difficulty.[75]

There was no question, however, of any negotiations with the Poles in St. Petersburg when the British Ambassador received the dispatch concerning Poland. On its receipt he began by getting in touch with his French colleague. Asked about his instructions with regard to Poland Mortemart replied that they were of a general character and intended rather for the moment of Polish defeat – "which he did not think by any means arrived", added Lord Heytesbury; the Netherlands Minister, who knew of the intercourse between the two Ambassadors, noted that these instructions, written in Paris after the news of the battle of 25th February, were sent "avec telle précipitation que le Cte Pozzo di Borgo n'a eu que le tems d'écrire quelques lignes de sa main par cette occasion." It was thought natural, therefore, that Mortemart made little use of these instructions, though in view of Heytesbury's comment it might have seemed strange that he only expressed the wish "qu'après l'entière soumission du Royaume les formes extérieures ne fussent pas trop blessées", so as to avoid popular indignation in Paris and the troubles which would result from it for the French Government. Nevertheless, the British Ambassador considered it to be his duty to see Nesselrode on the subject of Poland. The reception was cold; he learnt that Mortemart did not touch upon the question of the Polish provinces incorporated with Russia – the French Ambassador later confirmed this to him – and in all other matters the British overture met with no response. It could not, however, be qualified as an entire failure: Mortemart spoke on Poland, so did Heytesbury. "The apparent understanding between England and France upon this subject", concluded the British Ambassador, "is a much greater source of regret to this [Russian] Government than anything contained in the representations I was called upon to make." [76]

[75] Cf. Leslie, p. 168; Nicholas to Constantine, 3 Apr.: "Nous avons reçu de nouvelles du plus haut intérêt par Berlin..." if the constitution be maintained "il [Skrzynecki] répondait à l'instant de la soumission de l'armée. Que Czartoryski voulait venir ici en députation pour porter soumission, etc" Correspondance, p. 149. Skrzynecki was neither authorized to make such a proposal nor, probably, did he mean it seriously, cf. Schiemann, *Geschichte Russlands,* III, 105; Mortier to Sebastiani, 8 Apr., A.é. Prusse. This dispatch, however, was written after the news of the Polish successes, cf. below, p. 132.

[76] Heytesbury to Palmerston, 13 Apr., F.O. Russia; Van Heeckeren to Verstolk, 2 Apr., R.A. For Palmerston's instructions see above p. 129.

There were no doubt other grounds to strengthen the regret felt by the Russian Cabinet on account of the interest of the two Western Powers in Poland. After the Polish offensive move and two minor victories at the very end of March, there came in April the news of the rising in Lithuania, and, in the second half of this month, of a major success at Iganie, some 50 miles East of Warsaw. Within a fortnight all reports concentrated upon Lithuanian events. Great alarm was raised in Berlin and Vienna. Bernstorff told the British Minister that he would have preferred an actual and general war to a rising in the rear of the Russian army. Baron Van Heeckeren from St. Petersburg foresaw "une impression vive en Europe". Mortemart at first belittled the movement – and this opinion coincided in time with his attitude towards the approach of Heytesbury described above – but before the end of the month he came to recognize its increasing importance. It was reported in Berlin and in Vienna that Lithuanian soldiers of the Russian army joined the Poles in bands of 200 to 300 men. On 19th April the British Ambassador thought that the myth of Russian power had been completely destroyed and reported that all available Russian forces were employed, while in a subsequent dispatch he expressed his belief that "a long period is now likely to elapse before any decisive blow is struck." [77]

Before long echoes of the Polish offensive were felt beyond the Russian Empire. The French representative in Berlin found Bernstorff's opinions on European affairs very moderate and attributed this chiefly to the "avantages éclatants" of the Poles. As mentioned above, he reported the alleged negotiations in Poland. As a request for assistance was expected to arrive from St. Petersburg in the Prussian capital Mortier expressed his conviction that it would not be complied with, because the Berlin Government wanted to exercise some influence in the Belgian affairs. Only ten days after Mortier had reported that the Court and diplomatic circles had repeatedly given evidence of high spirits caused by the [false] news of a revolution in Belgium in favour of the Prince of Orange, he heard from Ancillon that Prussia would consider the candidature of

[77] Chad to Palmerston, 6, 18 Apr., F.O. Prussia; Mortemart to Sebastiani, 9, 15, 23 Apr., A.é. Russie; Van Heeckeren to Verstolk, 6 Apr., Van Spaen to Verstolk, 13 Apr., R.A.; Heytesbury to Palmerston, 19, 20 Apr., F.O. Russia; Chad gave credence to a report of the Danzig consul that Wilno and Kovno were taken by the insurgents, cf. to Palmerston, 19 Apr., F.O. Prussia.

Prince Leopold, "la maison d'Orange n'ayant plus de chances pour régner en Belgique." At the same time it was believed in Berlin that if Diebitsch suffered one more defeat, no new forces would be available on the Vistula before the autumn, and even that, if he was unable to win a major battle within a few weeks, the year's campaign would be finished. Uneasiness was increased by Nicholas' ukase concerning the Lithuanian insurgents, "digne des siècles de barbarie", to quote Mortier, which produced a bad impression upon the Prussian people and thus became "un sujet de vive anxiété et de nombreux embarras pour la cour de Prusse: ses agens ne le dissimulent pas", commented the French *Chargé d'affaires*.[78]

Less discouraging for the Eastern Courts was Chad's report at the end of April. He had spoken to Prince Anthony Radziwill, brother-in-law of the King of Prussia, and a brother of Prince Michael, the Polish Commander-in-Chief up to the battle of 25th February. The Prince considered General Skrzynecki, the new Polish commander, as a Royalist and added – as he had always maintained, commented Chad – that the revolution was now being carried on by others than those who were responsible for its outbreak. The Polish nobility had thus been driven into battle, "with no expectation of ultimate success, and no other object but to avoid dishonour." The Prince hoped that the Emperor Nicholas would make some concessions, but did not conceal that the Lithuanian insurrection rendered this much more difficult.[79]

Anxiety was also felt in Vienna, the more so as reports from Poland coincided in time with those from Constantinople about the French Ambassador's *démarche* to the Turkish Government with regard to the alleged immediacy of war. Already after the very first news from Poland in April Metternich feared its influence upon the other Polish provinces; shortly afterwards he told Cowley that Russia's strenght had at all times been exaggerated. Russia now could bring no more than 40 to 50 thousand men against France; this remark was probably made in view of the development of Italian affairs, the main concern of Austrian foreign policy.[80]

Lastly, Paris was not slow in realizing that Polish affairs "donnent une nouvelle face à la situation générale". According to the Nether-

[78] Mortier to Sebastiani, 10, 18, 23 Apr., A.é. Prusse.
[79] Chad to Palmerston, 27 Apr., F.O. Prussia.
[80] Maison to Sebastiani, 7 Apr., A.é. Autriche; Cowley to Palmerston, 9, 12, 13 Apr., F.O. Austria. For the Constantinople affair cf. below, p. 142.

lands Minister, the stock exchange and the Parliament as well as the Ministry were very much engaged with the Polish news. There were fears that the *parti du mouvement* would receive a fresh stimulus. The general excitement could not fail to impress the British Ambassador; he reported, however, that military spokesmen, including Sebastiani, did not find any material change in the position of the Poles.[81]

They had had hardly any success so far in moving the French Government into any diplomatic action in their favour. Following on the activity of a number of more or less self-appointed agents, in the beginning of March an official Polish mission, composed of Count Plater, who came from Poland, and an old Napoleonic soldier, General Kniaziewicz, resident at Dresden since the fall of the Empire, arrived in Paris. Their arrival coincided with the news of the battle of 25th February and Sebastiani received them coldly; there could be no question of presenting credentials. But in view of the popularity of the Polish cause they obtained a certain, almost semi-official, position in Paris and the French Ministers could not very well refuse to receive them. In the course of half a year of activity they had a number of interviews with Sebastiani, and presented various memoranda and circulars written either in Paris or in Warsaw. Some of these documents appeared in French and English newspapers.[82]

The first steps of the Polish envoys in Paris were, from the diplomatic point of view, quite discouraging. Not only was the news from Poland bad in the beginning, but also the Italian crisis engaged the attention of the French Cabinet, while the intractability of the Belgians presented another source of trouble. It was only when, to the envoys at least, Italian and Belgian complications seemed to be drawing to a close that at the end of March they went to see the Minister for Foreign Affairs with a hope of obtaining a pos-

[81] Duc de Dalberg to Talleyrand, 12 Apr., Talleyrand, *Memoirs,* IV, 140 Fagel to Verstolk, 11 Apr., R.A.; Granville to Palmerston, 11 Apr., F.O. France.

[82] Cf. Dutkiewicz, p. 78-9. The MSS of the Bibliothèque Polonaise in Paris contain a wealth of material concerning the diplomatic activity of the Paris mission, as well as of that of Wielopolski and Walewski in London and of other agents. Several of the documents confirm the semi-official status of the Polish representation. There are letters of recommendation for officers who went to Poland, mostly from General Pelet, Chief of Staff of the Minister for War, Marshal Soult which request an introduction "auprès du Gouvernement de Pologne". Most letters are addressed to MM. le Cte Plater et le Général Kniaziewicz, Envoyés de Pologne.

itive statement. Sebastiani had just received Mortemart's dispatch of 9th March, and tried to explain to the Poles the advisability of submission. He said that Nicholas was more than ever ready to enter into negotiations. This was far from being the object of the mission; the Poles would only agree to treat with Nicholas if there was mediation of some kind on the part of the Powers. The subsequent interview with Sebastiani, held after the news of the Lithuanian rising, brought equally negligible results. According to Sebastiani, Nicholas would not hear of any mediation whilst the French Government possessed no means of exercising any pressure upon Russia since Belgian affairs were about to be arranged. This suggestion was new and the envoys do not seem to have been *au courant* of the proceedings with regard to Belgium, since they acquiesced in Sebastiani's statement regarding the adjournment of the Luxemburg settlement for two months, but they came to the conclusion that as long as the affairs of Belgium remained unsettled, the French Government was able to influence those in Poland.[83] As things were, however, the contrary appeared more and more to be the case.

V

On 10th April the French representative in Brussels, General Belliard, arrived suddenly in Paris. Belliard not only found himself unable to make the Belgians agree to the "Bases de séparation", but was convinced of the necessity of a re-adjustment of policy with regard to Belgium. His arrival coincided with the reports of major changes in Poland.

Sebastiani gladly seized Belliard's sudden appearance in Paris to review the Belgian question once more at the British Embassy. He pointed out at first one of the Belgian demands, Zealand-Flanders. Then he suggested, as a measure of stability for Belgium, the election of a king prior to any territorial settlement. But Granville maintained that, failing the agreement on the Belgian side, the Federal forces should occupy Luxemburg before anything else was done. The French Minister protested against the war – as it seemed to him – being carried on so close to the French frontiers: Luxemburg could easily be defended. The British Ambassador merely answered that this difficulty might be removed by increasing the

[83] The Paris envoys to the Polish Government, 31 Mar., 15 Apr., B.P.

force to be employed for the purpose. The interview becoming unpleasant, and the Minister running out of arguments, he remarked at last that, after all, it would have been possible to make the Belgians submit to the decision of the Conference, had there been frank co-operation between the French and British representatives in Brussels; now it could only be achieved by removing Ponsonby who had been compromised through his avowed Orangist sympathies.

It has been shown that, only a few days earlier, the Orange candidature made Sebastiani hold out the menace of armed French intervention; nor was this the first time that the French Government had requested Ponsonby's withdrawal. Palmerston, however, hardly took into consideration any of the proposition contained in the extensive report from Paris; he pointed out the circumstances outside Belgium, not in Belgium, as the origin of the new French projects. "And first I should say that I think I see in Sebastiani's tone the rainbow of Skrzynecky's victories", began his answer to Granville. "But however the French may raise their pretensions in consequence of the reverses of the Russians, still we ought to be steady to our point, since we have never required more than is strictly just; and events in Poland cannot make it less so than it was. If the French are bent upon encroachment and war, no concession will keep the peace." Other foreign observers, in London and Paris, attributed a similar influence to the Polish events.[84]

The Paris report had not come to the Foreign Secretary as a surprise. Some days before he had had the opportunity to expound his views on France again, when the former Polish agent, Wielopolski, and the newly arrived Walewski approached him and obtained another interview. He repeated to them what he had told Walewski a week before. There was, in his opinion, no possibility of intervening in any way whatever to obtain the independence of Poland; the Poles could only ask for a status quo of the Kingdom, if there was a guarantee that in the case of foreign intervention they would submit. To this Walewski offered the Polish interpretation of the Treaty of Vienna, from which the Poles drew the conclusion that the provinces incorporated in Russia had also been subject to this agreement.

[84] Granville to Palmerston, 11 Apr., F.O. France, Palmerston to Granville, 13 Apr., Bulwer, II, 70, cf. above, p. 127; Matuszewic to Mme Lieven, 13 Apr.: "Les succès partiels des Polonais ... portent beaucoup de mal ... la résistance des Polonais influe sur le langage que la France tient dans les affaires de l'Italie et sur les nouvelles intrigues, qu'elle renoue en Belgique." Lie MSS; James Rothschild to Salomon Rothschild, 13 Apr., Corti, p. 29.

Palmerston took some notes and promised to examine the case more closely, but then immediately referred to his own difficulties. The British Government was not entirely free to act abroad, because of internal complications, but above all because of French ambitions it was essential that it should continue on the best possible terms with Prussia, Austria and Russia. "Dans toute autre circonstance", continued the Foreign Secretary, "l'Angleterre aurait tout fait pour voir le rétablissement de la Pologne", this being necessary for European equilibrium. But now, "ce n'est qu'en lui représentant les choses sous le point de vue de donner de la force à la Russie, que l'on pourrait engager l'Angleterre à faire quelque chose." This suggestion did not, of course, exclude the possibility of an intervention, but by no means met the standpoint of the Poles, especially as it almost excluded the possibility of France participating in such proceedings. "Si les affaires de Belgique étaient tout à fait terminées", Palmerston went on, "nous pourrions, peut-être, entrer plus facilement dans la combinaison avec l'Autriche; mais dans ce moment-ci nos plus grandes craintes sont du côté de la Belgique, et nous aimerions mieux renforcer la Prussie que l'Autriche." At this point the Poles could not go on. Walewski tried to get something done through Lord Holland and visited him as well but from that interview he only gained the impression of "beaucoup de sentiments mais... un bien petit résultat." [85]

Lord Holland, however, took a fairly lively interest in Polish affairs, and at this very time was trying to introduce this motive in British foreign policy; both Grey and Granville were approached. Lord Holland more than any other member of the Cabinet believed in the sincerity of the French Government; he trusted Talleyrand. It was to bolster the prestige of the French Government that British influence should be used; were it not constantly made to suffer humiliation in the estimation of the public, its efforts to obtain compensation elsewhere would cease. Thus British mediation could reconcile France and Austria in Italy, where neither party desired war, and it could equally bring them together in their common interest, which was to prevent the annexation of Poland by Russia. And this was in the British interest as well: "Any successful interposition by treaty and remonstrance [for the Poles]", he wrote to Grey, "would reconcile the public opinion of all countries and contribute

[85] Walewski to the Polish Government, 8 Apr., to Plater, 31 Mar., B.P.

not a little to pacific spirit in England and France." On the contrary, though the fall of Poland, "France will be covered by infamy & some portion of her disgrace would fall upon us." But Polish affairs had not been sufficiently explored by British foreign policy because, thought Lord Holland, Palmerston was more accessible to the suggestions of Princess Lieven – unfortunately – "for it is clearly the policy of Russia, if not to embroil this side of Europe in actual war, to prevent any cordial co-operation and success between France and England". Also at this time Lord Holland was enquiring about Polish affairs from Sir Robert Adair, whose diplomatic experience included a mission to Vienna in 1806.[86]

On the occasion of a subsequent letter to Grey in April Lord Holland stressed the unmistakably warlike intentions of Russia, of which he said there was enough evidence in her belated recognition of Louis Philippe, the Grand Duke Constantine's papers found at Warsaw, "and the unwillingness to acknowledge anything but a Prince of the House of Nassau" in Belgium; whilst now, through her declaration of support for Austria, she was pushing the latter to extreme measures in Italy. "I should hope that the events in Poland may have... checked the ardour of Russia for any interference which would call for exertion from her", he commented, "but they will not have diminished her wish to see the West of Europe embroiled, if not in a war, in disputes and jealousies that will prevent them from taking much interest in the concerns of that country." [87]

There is no evidence that Palmerston was, in his Polish policy, guilty of partiality caused by the influence of Mme Lieven. But the public news from France supported the belief that the revolution in Poland had been directed from France – suffice it to mention the incident during Lafayette's speech referred to above. Whether sincerely or not Palmerston told both Walewski and Wielopolski that he considered the Polish movement to have originated in France. On their replying in the negative he observed that the

[86] Holland to Granville, 8 Apr., Granv MSS; Holland to Grey, 7 Apr., Adair to Holland, 13 Apr., an enclosure to Adair's letter to Grey of 27 May (cf. below, p. 151), Howick MSS; Holland's letter to Grey contains the following passage: "Those damned Belgians are the origin of all mischief. I heartily wish they had been well dismembered & partitioned between France, Holland, Prussia and England 16 years ago & when the time comes as it inevitably will, if I am alive, I shall rejoice at it." This passage proves not only the relative impopularity of the Belgian policy in England but also the belief in the fragility of the independence of Belgium, cf. below, p. 160.

[87] Holland to Grey, 12 Apr., Howick MSS.

Poles in France were greatly embittered against France for her exhortations as regards the rising, followed by a total abandonment when it came. [88]

At this time, within a week from the first news about the Lithuanian rising, Russian difficulties received considerable publicity in London. Like French military circles, Wellington was reluctant to attach significance to the Polish victories; they had been made possible by the dispersion of the Russian forces necessitated by lack of provisions. But he believed that the rising in Lithuania had changed the situation. As to the general public, Princess Lieven observed that, fortunately, Polish affairs "se voyent au milieu de la réforme," but she herself found the news lamentable. Incidentally, on the same day Palmerston wrote to instruct Heytesbury on the subject of mediation. In the true interest of Russia a mediation which would bring an end to the struggle should, he thought, be welcomed by the Imperial Government. The Foreign Secretary noted that even a victory would not be entirely sufficient to suppress the Poles, that Prussia and Austria were unlikely to give assistance against them, for fear of France. The emperor Nicholas should therefore be more inclined towards a reconciliation.[89]

There seems to have existed a good climate for the Polish cause in London. Not only Lord Holland and Adair, but also Granville and Palmerston did not conceal their Polish sympathies. As for the rest of the Ministers, Walewski reported his daily contact with them, whilst Mme Lieven found that he was "recherché dans la société". Unlike Wielopolski's arrival in January, that of Walewski had not been publicized by the newspapers, and this made it easier for the Ministers to see him; even Wessenberg, one of the Austrian delegates communicated with him. All this, however, failed to bring about a change in official policy. Palmerston had, however, other arguments besides France only. At a subsequent meeting with Walewski, on 21st April, Palmerston referred to Sebastiani's opinion that the Polish victories were by no means decisive. "Je me contentai de lui répondre que le général Sebastiani est aussi mauvais général que pauvre diplomate," reported the Polish envoy. Palmerston seeming-

[88] See above, note 85.

[89] Wellington to Alava, 13 Apr., Wellington, *Suppl. Dispatches,* New Series, VII, 430-2; Palmerston to Heytesbury, 19 Apr., Heytesbury MSS; Mme Lieven to Nesselrode, 19 Apr., "J'ai tous les Dwernickis, les Uminskis, les Skrzyneckis sur la poitrine, ils m'étouffent." Lie MSS.

ly concurred with this opinion and assured Walewski that all English military men as well as he himself considered the successes decisive for the campaign, "que pourtant le moment n'était pas encore venu d'intervenir hautement dans les affaires de la Pologne." In spite of this disappointing statement the pamphlet published by the Polish envoy, which stated the case for intervention, enjoyed, according to Walewski, "assentiment" of the Cabinet, while the Austrian mission in London requested several copies of this paper for the purpose of sending them to Vienna.[90]

Palmerston nevertheless was active in other ways to the advantage of the Poles. He warned Bülow that if Prussia intervened in Poland and French troops entered the Rhineland, no assistance would be given to Prussia by England. He also hinted to Esterhazy, the Austrian Ambassador, that a Russian application for military aid would afford Austria a fair opportunity to offer mediation instead. "There will be no harm in hinting this to Apponyi, "Palmerston informed the British Ambassador in Paris, "though I am glad to find out that Austria has declined the aid which Russia has applied for." [91]

Palmerston seems to have known only of the rejection of the Russian request, but Grey told Mme Lieven that there had been some offer of mediation by Austria. Information of an attempt at reconciliation through Vienna came to the knowledge of the Netherlands Minister there. Grey showed no little astonishment at what he called a "mismanagement" of things by Diebitsch, and speaking of the supposed Austrian offer he observed to Mme Lieven "qu'il faudra bien finir par s'en mêler." For the rest, the Princess had no good news to record at all; hardly a week had passed since the Polish news had been so regretfully interesting only to herself, when she noted that Skrzynecki's bulletin from Siedlce made even the King "m'en parler avec chagrin." [92]

[90] Palmerston to Granville, 29 Mar., 18 Apr., Bulwer, I, 49, 73; Granville to Palmerston, 15 Apr., Palm MSS; Walewski to the Polish Government, 22 Apr., B.P.

[91] Palmerston to Granville, 22 Apr., Bulwer, II, 76. No reference to a Russian request is found in J. Dutkiewicz, *Austria wobec Powstania Listopadowego* (Austria and the November Rising), 1933, except the Austrian refusal to occupy the Republic of Cracow, p. 72.

[92] Mme Lieven to Alex. Benckendorff, 22 Apr., to Nesselrode, 26 Apr., Lie MSS; Van Spaen to Verstolk, 23 Apr., R.A.; Grey to Mme Lieven, 13 Apr., 19 Apr., *The Lieven–Grey Correspondence,* II, 109, 212. Siedlce was the nearest major town in the neighbourhood of Iganie, cf. above, p. 132.

In the meantime the first Reform crisis had been overcome by a vote taken just at that time which, in addition to the military news, seemed to create better prospects for Walewski's representations. On 24th April Lords Grey and Durham put to the Polish envoy the straightforward question which was to be expected: what was France going to do for the Poles? This was a hard question to answer. Walewski seemed to be getting on well with Talleyrand, who had assured him that Polish affairs would be settled within two months; but Walewski possessed no official information to confirm this. The French Cabinet had informed the Prussian *Chargé d'affaires* that France would not suffer a Prussian intervention in Poland; and if Prussia considered herself bound to it by previous declarations, the French Government could only reply with their counter-statement that in such a case they would order French troops to march not only into the Rhineland, but further afield. This, however, Walewski heard only as 'bruits'. He believed that Heytesbury and Mortemart were acting jointly; it has already been shown that there had hardly been anything of the sort. Lastly, he knew in what embarrassment the French Government found itself by the unfortunate step of General Guilleminot at Constantinople, not the least as he himself could observe, with regard to the British Cabinet. He had, therefore, to leave the question virtually unanswered: he merely said that the French Government would do nothing except in agreement with London, and of this intention the British Ministers possessed only too much evidence.

The opportunity to push the matter of intervention may well have remained unexploited, but Walewski was able hopefully to remark in his report that he knew on good authority that Palmerston was being particularly occupied with Polish affairs.[93]

If this was the result of the Russian misfortunes on the Vistula, Polish events also had repercussions elsewhere in Europe.

[93] Walewski to the Paris envoys, 25 Apr., the Paris envoys to Walewski, 25 Apr., B.P. The French declaration was no secret, nor were the public views of another aspect of the new situation in Poland; cf. Fabricius to Verstolk, 22 Apr., R.A.: "Les succès des Polonais auxquels on croit devoir ajouter foi ici, ont fait naître des inquiétudes en tant qu'on croit que la Prusse et peut-être, l'Autriche même, pourront difficilement s'empêcher d'intervenir pour s'opposer à leurs progrès, s'ils deviennent trop décisifs."

CHAPTER IV

THE EIGHTEEN ARTICLES AND THE INTERVENTION IN POLAND

During the month of April there was a new development in Belgian affairs: not only did the Belgians remain untractable, but the French Government seemed also to take a stronger line in their favour. For Palmerston, as has been seen, this was a direct consequence of the Polish events. So it was understood in Brussels where a tendency could be observed to underline the change. Early in April the Belgian envoy in Paris took a gloomy view of the situation, as he described the weakness of France, her exhausted finances and the unpreparedness of her army; this coincided with Sebastiani's exhortations to the Provisional Government to accept the conditions laid down by the Conference. At the beginning of May, however, Count Le Hon reported the "general opinion" in the French capital that, owing to the paralysing influence of the Polish victories upon the policies of Russia, Austria and Prussia, "the circumstances were never so favourable for our establishing a Belgian State." [1]

The French Government may not have wished to avow a sudden change in their attitude, but suspicions were soon raised as to their sincerity. Thus the Guilleminot affair furnished further evidence of poor understanding between London and Paris.

On 19th March the French Ambassador in Constantinople, Count Guilleminot, inspired by the private letter from Marshal Maison quoted at the beginning of Chapter III, submitted a note to the Turkish Government in which he urged it to prepare for the approaching war, in which France, supported by England, would fight against Austria, Russia and Prussia. Within a few days the matter became known in the Turkish capital, and some weeks later in those of other European states. The Ambassador was recalled. It is unlikely that he had had any instructions which would warrant his *démarche:* these would have had to leave Paris in February, at a

[1] Le Hon to Surlet de Chokier, 2 Apr., Th. Juste, *Le comte Le Hon,* 1867, p. 44; to Lebeau, 9 May, *Memoirs of Leopold I,* ed. Th. Juste, English translation 1868, I, 139.

time when the French Government was on the defensive. When Granville, however, wanted explanations of this affair, Sebastiani put himself out to prove the absence of any relevant instructions to Guilleminot, but he only earned the following observation: "I did not feel myself called upon to question the validity of the proofs he adduced nor did I intimate that separate, private and confidential instructions might be given to an Ambassador." [2]

Another incident which stirred French public opinion at a later time showed that in spite of Mortemart's efforts, there was little hope of better relations between St. Petersburg and Paris. An article appeared in the "Journal de St. Pétersbourg" on 28th May which attributed to Nicholas the intention of combating French liberalism. When in the second half of June this news arrived in Paris, public indignation was so great that Périer invited the Russian Ambassador to an urgent interview. Pozzo's representations were not enough for him. The Imperial Cabinet was asked to furnish an official explanation, but before it could arrive the publicity in the matter exercised no little influence in public affairs.[3]

I

If the French Government in applying a new line in Belgian affairs remained within the limits created by previous developments, the Belgians, perhaps encouraged also by the self-appointed mission of Belliard to Paris, inclined to disregard the work of the Conference. Thus the Regent himself not only believed in a new treatment of Belgian affairs after the Polish successes, but expressed his conviction that the Belgians would now feel relieved of any special gratitude towards France for her new policy, as they would consider it to be a matter of course in consequence of the events in Poland. Similarly, the Belgian Envoy in Paris approached Granville, trying to impress on him the necessity of electing a king for Belgium, prior to territorial arrangements (ensuing from the acceptance of the Protocol of 20th January). But the Ambassador suggested that the Powers would not shrink from using force in order to introduce those arrangements, and at the same time France acceded

[2] Dutkiewicz, pp. 84-5; Granville to Palmerston, 22 Apr., F.O. France, cf. above, p. 98.

[3] de Guichen, p. 383.

to the "Bases", having been previously bound to them only by the signature of Talleyrand.[4]

Accordingly a Protocol (No. 22) was drawn up by the Conference on 17th April which stipulated, that if the Belgians should not agree to the "Bases de séparation", coercive measures would be applied to them, whilst the independence of Belgium was stated to be provisional and only to be recognized after the acceptance of the Bases. The British Ambassador in Paris at this time could not but regret that Louis Philippe's address to Parliament did not mention the provisional character of the Belgian independence.[5]

When the Protocol arrived in Paris, Sebastiani summoned a meeting of the representatives of the Conference Powers. He succeeded in making them accept a resolution by which French accession to the Protocol was registered and the date of 1st June was set up for the Belgians to comply; this, however, was coupled with the understanding that the occupation of Luxemburg would also now be considered as a coercive measure, subject to the approval of the Conference. The French Minister also tried unsuccessfully to suggest that this occupation should be associated with the French entry into Bouillon. "Some surprise was expressed by my colleagues and myself at the novelty of the proposition", commented the British Ambassador.[6]

Talleyrand tried to have the Paris resolution of 22nd April accepted by the Conference as well, which would make Luxemburg safe at least until 1st June, but Palmerston, pretending to be in a hurry to leave London for his Cambridge elections, prevailed in having discussion on this subject adjourned. There was, therefore, no interference on the part of the Conference, and the progress of preparations connected with the military occupation of Luxemburg caused Louis Philippe to complain anxiously.[7]

Soon, however, the next Protocol, no. 23, of 10th May, was drawn up in such a way as at last to break down Belgian opposition. It stipulated that the Belgians would have to accept the "Bases de séparation" before the 1st of June, and that this acceptance would

[4] Sol to Belliard, 10 Apr., Belliard, *Memoirs,* II, 227-8; Granville to Palmerston, 15 Apr., F.O. France.

[5] Granville to Palmerston, 20 Apr., F.O. France.

[6] Granville to Palmerston, 22 Apr., F.O. France.; cf. Franqué, pp. 85-6.

[7] Bülow to Frederick William III, 29 Apr., Francqué, p. 101; Louis Philippe to Sebastiani, 3 May, *Revue des Deux Mondes,* LVI, 304.

imply taking preliminary steps to evacuate the Grand Duchy as well; should that not be done, the Conference proposed suspending their relations with the Brussels authorities on the 1st of June and agreeing to the need of military measures as contemplated by the Federal German Diet. An attempt at a renewal of hostilities would be considered as directed against all the Five Powers. But the German Diet would then still be requested to furnish information about the number of troops and their proposed action. This amounted to a camouflaged acceptance of the Paris resolution of 22nd April, in accordance with the preamble to the Protocol which affirmed the necessity of some action but equally stressed the consideration of the general peace.[8]

This Protocol increased the tension regarding Belgian affairs. Sebastiani, with some reluctance, seemed to consider it as the prelude to a final arrangement though he seized the occasion to hint to Granville the difficulty of Luxemburg remaining in the German Confederation. Palmerston also believed that there would be a speedy end to the negotiations. But the contesting parties took an entirely different attitude. The Hague regretted the lack of sanctions which would guarantee the execution of the Protocol provisions. Minister Verstolk attributed this omission to English internal difficulties and to the development of the campaign in Poland. In Brussels Government circles thought the publication of the Protocol impossible. Ponsonby decided to go to London and explain the situation. He arrived at the seat of the Conference to the great surprise of Palmerston to report in person that Belgium was on the verge of another revolution, the establishment of a Republic and war with Holland.[9]

Palmerston at first thought that the Belgians could be induced to agree to the Protocol, if it were explained to them that the Confederation might have the right to turn them out of Luxemburg, but would never do so. According to him the Confederation would not do the job unless paid by the King of the Netherlands, who had no money for that purpose, so there was no danger to the Belgians of

[8] The Protocol in B.F.S.P., XVIII and in *Papers Relative to the Affairs of Belgium,* State Papers, vol. XLII A.

[9] Sebastiani to Mortier, 17 May, A.é. Prusse; Granville to Palmerston, 20 May, F.O. France; Verstolk to Falck, Van Zuylen, 16 May, R.A.; Palmerston to Bagot, 16, 17 May, Bagot MSS; Bülow to Frederick William III, 10 May, Franqué, p. 101.

such an action being taken. But he soon inclined to the view that the acceptance of the Belgian crown by Prince Leopold was the only means to avoid the extremes expected by Ponsonby. He was then temporarily relieved by Van Zuylen's information that King William I would negotiate about the cession of Luxemburg – this being the condition laid down by Leopold – but had to learn shortly afterwards that that was not the case.[10]

In the meantime, however, on the strength of Van Zuylen's advice as well as of Ponsonby's persuasions a new Protocol was signed which constituted a first breach in the, now of long standing, "Bases de Séparation". On the 21st of May the Conference for the second time appointed for the Belgians the 1st of June as the ultimate date for acceptance of the "Bases", but at the same time held out to them the hope that the King of the Netherlands might relinquish his possession of Luxemburg. "The advantages obtained by the Protocol," commented Palmerston, "are that the Russians have for the first time agreed to express a wish that Leopold might accept... and further, that the Five Powers engage to request the Confederation to delay the march of their troops pending the negotiation about to be opened for an exchange for Luxemburg." But the Protocol was not to be communicated officially to the Belgian Government.[11]

The Netherlands Government, however, began to lose its patience. The Minister for Foreign Affairs told the British Ambassador that the state of affairs was becoming intolerable and that the King might be led to make a bilateral agreement with France. This would result in what Palmerston feared: the entry of the French and Prussian troops into Belgium and the consequent inevitable partition of that country. Bagot promptly suspected some secret talks going on between The Hague and Paris.[12]

The 1st of June had passed but on the Belgian side there was no sign of any conciliatory disposition; on the contrary, on the basis of Ponsonby's information, Palmerston thought it necessary to warn the Netherlands Government of the likelihood of the Belgians commencing hostilities. On the other hand, it eventually appeared that Lord Ponsonby had not considered it advisable to communicate

[10] Palmerston to Bagot, 20, 22 24 May, Bagot MSS; cf. to Grey, 15 May, Howick MSS and Franqué, p. 105.

[11] Palmerston to Grey, 21 May, Howick MSS.

[12] Bagot to Palmerston, 23, 24, 27 May, Bagot MSS.

the Protocol of 10th May, while he utilized the discretion left to him to interpret the not-communicable Protocol of the 21st to make the Belgians believe that the Conference would obtain Luxemburg for them.[13]

Ponsonby's conduct aroused indignation at the Conference and he was recalled (6th June), but the election of Prince Leopold had taken place. The Dutch plenipotentiaries for the second time notified Palmerston that after the expiration of the term set forth for the Belgians the King of the Netherlands considered himself free to employ any means he had at his disposal to terminate the question. Nevertheless, they tried also to make the British Government adopt a more sympathetic line with regard to Dutch interests.[14]

Palmerston did not expect that Leopold would in these circumstances accept the crown of Belgium, and viewed the situation in these terms: "The great point is to keep the Dutch from beginning hostilities, but also to prevent them bribing the Belgians to do so. It is said that the late riots in Antwerp were paid for in Dutch coin while the Brussels disturbances were purchased in five-francs pieces. The only doubt we can have is whether, if war begins, the French Government can keep his unruly subjects out of Belgium, and how many volunteers the Dutch would find in the Belgian ranks. I fear that whatever may be the improvements of the Dutch army, four or five thousand French officers and men, mixed up with the Belgians would make the matter serious for the Dutch, and more than that number might probably be expected." The Conference drew up a memorandum for the representatives of the Conference Powers resident in The Hague in order to warn the King of the Netherlands against a possible Belgian invasion after the suspension of relations with the Brussels authorities by the Conference.[15]

This was no less a problem for France. Sebastiani and Périer, through Talleyrand as well as through the British Ambassador, tried to impress upon London the difficult situation of the French Government in consequence of the development of Belgian affairs,

[13] Palmerston to Falck, 28 May, R.A.; Ponsonby to Lebeau, 27 May, Howick MSS.

[14] Bagot to Palmerston, 23, 24, 27 May, Bagot MSS; Falck, Van Zuylen to Verstolk, 4/5 June, to Lord Holland, 8 June, R.A.

[15] Palmerston to Bagot, 7 June, Bagot MSS; "Memorandum pour les Ministres des 5 Cours à La Haye", appended to Talleyrand's dispatch of 8 June, A.é. Angleterre.

which inevitably led it into coercive measures against the Provisional Government; Périer hastened to inform Granville that he considered Leopold "used up" through the belated disavowal of Ponsonby.[16]

These alarms were not left without evoking a response on the other side of the Channel. Lord Holland readily brought the fears of Granville to the knowledge of Grey who, however, seemed to have been already convinced: the French would not be mere onlookers in a Belgo-Dutch war. Moreover, he largely dismissed English obligations towards Holland: "... we now hear that Falck the organ of the Dutch Government wishes to embroil matters as much as he can, to defeat the combinaison Léopold..." If a war comes, "let the Dutch fight their own battle" and assist them only as far as to avoid the odium of coercing the Belgians falling on England. "This is evidently the object of the French Government and perhaps of others..."[17]

Palmerston, though primarily engaged in working out a formula of acceptance for Leopold in which the "Bases de séparation" would not have been abandoned altogether, answered, however, that the Belgian pretensions, notably to Maastricht, would lead to war, and a war in which Holland, backed by the Eastern Powers, would have to be defended by England as well.[18]

Talleyrand apparently was unaware of this divergence of views. At this time he noted that the difficulties originated from two causes: The Hague's intention to engage the Powers in a war, and Russia's aiming at focussing the attention of the Powers on affairs in the West, without any definite purpose except to see if she could not profit from it. The French Ambassador was daily seeing the English Minister and Prince Leopold as he expected to find there a similarity of views and interests.[19]

II

During May the Polish envoys in Paris continued their efforts to

[16] Granville to Palmerston, 8, 10 June, F.O. France, 10 June, Palm MSS; Sebastiani to Talleyrand, A.é. Angleterre.

[17] Granville to Holland, 8 June, Granv. MSS; Holland to Grey, 13 June, Grey to Palmerston, 11 June, Howick MSS.

[18] Palmerston to Grey, 13 June, Howick MSS.

[19] Talleyrand to Sebastiani, 13 June, Pallain, p. 406

engage the Ministry in a more active policy with regard to Poland. They were encouraged by the anxiety shown by Pozzo and Apponyi. Nothing, however, was achieved through Sebastiani and Périer, though Walewski believed in the former's favourable attitude. The French premier received General Kniaziewicz and read the Polish project concerning the request for mediation; while he considered it moderate, he regretted at the same time his inability to lay it before the Council. It was thought, however, that a better reception would be gained from Talleyrand.[20]

Grey, according to Walewski, "se tenait dans une réserve invulnérable." He could not acknowledge the communication from Prince Czartoryski, but Talleyrand told the Polish envoy that Grey had talked to him about that document. Lord Holland received it from Grey, and in his reply expressed his conviction that Poland could be "rescued" by common diplomatic steps by England and France. "Nothing but the persuasion in Russia, Austria and Prussia", he wrote, "that France and England never can agree to act in concert against the wishes of the two partitioning powers prevents Russia from declining..." such steps.[21]

Lord Holland observed Russian fear of such an Anglo-French agreement even in Heytesbury's dispatch though he considered the latter's presence at the Russian Court a major obstacle to a successful diplomatic action. But, above all, it was France whose foreign policy made such an action impossible. "We were all highly pleased with Casimir Périer", he wrote to Granville, "but wish he had a minister for foreign affairs with whom we had had fewer causes of difference and discontent... if we could agree cordially about all the matters in the west of Europe, we might by force of a drive & without risk of war do something for those noble fellows the Poles – Austria, once calmed about Italy, would follow our policy in that quarter..." [22]

Grey saw also Biré, a Frenchman who managed to get out of Warsaw and passed through Prussia pretending to be a British

[20] The Paris envoys to the Polish Government, 5 May, B.P.

[21] Walewski to the Polish Government, 5 May, B.P.; the letter referred to from Czartoryski is probably that of 15/21 March, in Howick MSS, brought by Evans (Walewski to the Polish Government, 2 May, B.P.) cf. below, p. 150; Lord Holland to Grey undated, but written probably about 10 May, Howick MSS.

[22] Lord Holland to Grey, same letter; to Granville, undated but written probably on May 14.

subject; somewhat unfortunately he was employed by the Warsaw authorities. In addition to the usual assurances that St. Petersburg would not hear of any mediation, Grey told him that perhaps a major victory would furnish a pretext for a cautious attempt at interference. He subsequently made the same statement to Walewski, adding that the English Government would, in such a case, take the initiative. "Il m'a répété ensuite et à plusieurs reprises: ces quelques jours seront bien importants pour vous", wrote the Polish envoy. "N'oubliez pas de venir me trouver aussitôt que vous aurez reçu une nouvelle positive." [23]

Palmerston admitted to Walewski that he would gladly have a consul at Warsaw to counterbalance the reports from St. Petersburg, but that such an appointment under the circumstances constitute an act of hostility towards Russia. Walewski re-opened the question some days later; with some reluctance Palmerston agreed to see a Mr. Evans, who had come to London at the expense of British subjects resident in Poland to seek for them the protection of the Government.[24]

Talleyrand counselled giving the greatest publicity to notes and circulars of the Polish Government. Palmerston read them; they appeared in many French and English newspapers. Talleyrand strongly recommended that Flahaut's mission to Berlin should be fully exploited, and Walewski, who was Flahaut's personal friend, came over to Paris to see him before his departure. The Ambassador also enlisted British support for Flahaut's mission, but the result was, as will be shown, rather meagre. Talleyrand told Walewski on the 23rd that he saw the termination of the Belgian negotiations approaching, which would enable him to concentrate on Polish affairs. He stressed the particular importance of Flahaut's mission: he should not be discouraged by first impressions. Flahaut was, according to Talleyrand, to propose mediation and the placing of the second son of Frederick William III on the Polish throne.[25]

If the Press was satisfactory, there seemed to be otherwise no good news for the Poles. Walewski, on his departure from London, saw "the horizon darkening" whilst Flahaut promised to spare no efforts, but planned "indirect action only". Kniaziewicz and Plater,

[23] Biré to Plater, 5 May; Walewski to the Paris envoys, 6 May, B.P.

[24] Walewski to the Polish Government, 5, 18 May, B.P.

[25] Walewski to the Paris envoys, 13, 15, 20 May, to Czartoryski, 30 May, B.P.

who, in the second half of May, each separately saw Sebastiani, obtained no positive statement except compliments and some vague advice to annul Nicholas' dethronement.[26]

Nevertheless, Sebastiani had addressed a dispatch to Mortemart concerning Poland. The French Ambassador was requested to propose mediation on the ground that the cholera epidemic was greatly assisted by the military operations in that country. The Minister also approached Granville to assure the British Government's support for this *démarche*. The British Ambassador had already had a conversation on Poland with Périer. The French Premier believed that both Austria and Prussia would protest against the incorporation of the Kingdom of Poland into Russia, "but he thought we – England and France – ought to endeavour to obtain real as well as nominal independence, or obtain at least some sort of security for the Poles against Russian oppression. He said that his account from Vienna indicated that even the Government there had a leaning to the Poles and no very friendly feeling towards the Russians, that they had connived at the passage of arms through their territory into Poland..." [27]

This was written on the 11th May and thus arrived in London at almost the same time as the news of General Dwernicki's defeat and his crossing into Austria, which came instead of the victory desired by Grey. The General's defeat also meant the collapse of the rising in Volhynia which was to have been organized by Dwernicki.[28]

Yet Grey's interest in the question continued; he at this time requested Sir Robert Adair for a copy of Adair's letter on Poland, written to Lord Holland in April. This letter developed Lord Holland's familiar point, common interest of all Powers in preventing Russia's aggrandizement. "It is fortunate", the letter said, "that the treaties of 1815, together with what is known to have

[26] Walewski to the Paris envoys, 20 May, to Czartoryski, 30 May; the Paris envoys to the Polish Government, 24 May, B.P. As for the Press, Fagel reported on the 13th the passing into opposition of the so far moderate 'Temps': "Mr Coste, son rédacteur, a déclaré à Mr C. Périer qui est son principal actionnaire, qu'il ne pouvoit pas continuer à écrire dans son sens, à moins que le Président ne voulût lui promettre d'aller à secours des Polonais." R.A.

[27] Sebastiani to Mortemart, 15 May, A.é. Russie; Granville to Palmerston, 16 May, F.O. France, 11 May, Palm MSS.

[28] J. Dutkiewicz, *Austria wobec powstania listopadowego* (Austria and the November Rising), 1933, pp. 60 sq; Palmerston to Grey, 14 May, on Lieven's information from Vienna dated the 3rd. Howick MSS.

passed on the subject of Poland at the Congress of Vienna, furnish the basis of a negotiation for such an object under the mediation of England." Austria, according to Adair, had already shown her distrust of Russia. "Added to this", the letter went on, "if you could get Esterhazy to speak out, you would find Austria *at this moment* very much alarmed and uneasy at large Russian forces in Poland and their possible increase." [29]

Unfortunately for the Poles, these suggestions were answered by the Austrian treatment of Dwernicki's corps. There were immediate requests from the Russian side to deliver the men as prisoners. The Austrian Government, placed in a difficulty, delivered only their arms and horses, refused the men, and, yielding to the Russian request not to let them free, put forward the condition that they should be maintained at the expense of Russia. Both the French and the English Governments prepared protests but these arrived in Vienna almost two months later, in considerably different circumstances.[30]

At the end of May Sebastiani invited the British Ambassador to a special interview concerned with Poland. From the positions of both armies there he considered that the general battle which was imminent had already been fought; whether gained or lost by the Russians, the battle would produce a crisis which France and England should avail themselves of in order to enter into discussions with St. Petersburg. Further he feared that a Polish victory might bring about Austrian intervention. But Granville, though not prepared to enter into the subject at any length, doubted the possibility of an interference – notwithstanding personal sympathy for Poland of the members of the Cabinet – and could only hold out the hope of the British Government trying to mitigate Russian severity in case of a defeat.[31]

Palmerston, however, saw Polish affairs in a brighter light. He believed Cowley's account making out that the prospect of a Russian success was very distant; he further observed that "if Skrzynecki succeeds in his movement upon Vilna, I shall look upon Russia as defeated." The Russian Embassy must have been uneasy about Palmerston's intentions; they noted with relief Matuszewic's

[29] Adair to Grey, 27 May, Howick MSS, cf. above, p. 138.

[30] Cowley to Palmerston, 7 May, F.O. Austria; Van Spaen to Verstolk, 14 May, R.A.; Webster, I, 188.

[31] Granville to Palmerston, 30 May, Palm MSS, 1 June, F.O. France.

report that the British Government had no plan of appointing a consul in Warsaw.[32]

The real or supposed intentions of France and England towards Poland received a considerable blow by the defeat of the main Polish army at Ostrolenka on 25th May. Yet the bad impression of this event disappeared much sooner than that of the battle of Grochów. Palmerston, Lansdowne and Holland expressed to Walewski their satisfaction that the issue of the battle was not so bad as it was represented by the newspapers; these were also influenced by the bulletin of the Polish mission in Paris, drawn up in a spirit of depression. Palmerston, reported Walewski "convient avec moi que le corps lancé en Lithuanie pourrait avoir un plus grand résultat que deux batailles gagnées." It is worth noting that the same point was raised at the time in Vienna. Lord Holland renewed assurances of the Polish sympathies of the whole Cabinet and stressed his particular point: common action of France and England "car ce n'est qu'en agissant de concert qu'ils peuvent vous être utiles". But Belgium he still considered to be an obstacle, and in any mediation which he was prepared to do his best to bring about, the Polish provinces incorporated with Russia, for which the Poles claimed special treatment, would present a major difficulty.[33]

Talleyrand was more encouraging. His reports from Paris confirmed that Walewski's journey had influenced the instructions given to Flahaut. The French Ambassador insisted on raising the spirits of the English public very much shocked by the battle of Ostrolenka. But, above all, he said, he hoped that the Belgian affairs were on the point of being arranged. "Ou ils consentiront aux protocols", he told the Polish envoy at the time of Ponsonby's recall, "alors, tout est dit, et si non, l'Angleterre et la France bloqueront l'Escaut, et il faudra bien qu'ils reviennent à la raison." Prussia would not be involved. But for Ostrolenka, he said, he would have begun his *démarches,* for which he had the fullest authorization; he would begin them, however, as soon as things quieted down a bit. Talleyrand sounded all the Ministers and found Grey the most difficult; he requested Walewski to try

[32] Palmerston to Granville, 3 June, Granv MSS. Matuszewic to Mme Lieven, 3 June, Lie MSS.

[33] Walewski to Czartoryski, 9 June, B.P.; Van Spaen to Verstolk, 13 June, R.A.

and influence him. Less official reports from the French Embassy said that Grey was "moins livré à l'influence Lieven". Accordingly, his attitude gave Madame Lieven matter for complaint.[34]

The Paris mission had so far achieved no success. Sebastiani had no positive statement to make to General Knaziewicz in their interview on 13th June except his "parole la plus sacrée" that the French Government was doing everything possible to help the Poles; in fact, he had written again on the subject both to Berlin and St. Petersburg but rather in the sense of preparing favourable conditions for submission. Talleyrand was requested to make the British Government agree to a joint protest to Vienna concerning Dwernicki's corps; Sebastiani's dispatch anticipated the Ambassador's own advice. Talleyrand reported previously that public opinion was against Austria, and now he ascribed this opinion to the Government as well but observed in his reply that no great efforts were to be expected, "parce que le Gouvernement anglais ne s'occupe jamais fortement que d'une affaire" and at present it is overburdened with two: the Reform and Belgium.[35]

The warlike feeling in France, Belgium and Holland, which greatly embarrassed the British Government at that time, seems to have been increased by division at home: the Dutch reports from London confirmed those by Bülow. The Cabinet was said to have been divided and the "as yet moderate" Ministers: Palmerston, Carlisle, Goderich and Grant, were on the point of resigning. Lord Grey took care to impress the Dutch plenipotentiaries with his views on popular feeling in France.[36]

Bülow himself went to great lengths to explain his motives in making the representatives of the Eastern Powers remain at the Conference after 1st June though no coercive measures with regard to the Belgians had been contemplated. Actually in spite of Ponsonby's recall a new Belgian delegation came to London on

[34] Walewski to Czartoryski, 9 June B.P. Mme Dino to Mme Adelaide 9 June, *Nouvelle Revue Rétrospective*, (1901) XV, 404; Mme Lieven to Nesselrode, 10 June: "J'ai vu Lord Grey hier ici... Je l'ai trouvé faible, poltron, aigri, voulant toujours nous plaire, mais plaignant, admirant les Polonais, poussant des voeux pour eux, cependant désirant que nous terminions aussi vite que possible; il regarde la bataille d'Ostrolenka comme excessivement importante, il espère que Diebitsch ne la laissera pas demeurer stérile dans ses conséquences." Lie MSS.

[35] The Paris envoys to the Polish Government, 13 June, B.P.; Sebastiani to Mortemart, 9 June, to Flahaut, 6 June, to Talleyrand, 13 June, A.é. Russie, Prusse, Angleterre; Talleyrand to Sebastiani, 10, 15 June, Pallain, pp. 410 sq.

[36] Van Zuylen to Verstolk, 14, 18 June, R.A.

8th June to offer the Belgian crown to Prince Leopold. According to Bülow the result of a withdrawal would be no less than Durham taking over the Foreign Secretaryship from Palmerston, an immediate recognition of Leopold as King of Belgium by England and France, and a considerable change of the attitude of the Western Powers with regard to Poland. This report, however, crossed on its way to London a dispatch from Berlin in which Bülow's accession to the idea of a bargain about Luxemburg was severely censured; in his answer Bülow merely pointed out that at the time of the signing of Protocol No. 23 it had been known in London that the Federal Diet's recess made military measures impossible for months.[37]

The provisional Government of Belgium realized that their country was unlikely to profit by a renewal of hostilities, and tried to prepare the ground for the acceptance of the new arrangement which was being negotiated. One of the members of the Belgian delegation in London, Count de Merode, invited Walewski to support their efforts. The popularity of the Polish cause in Belgium might warrant submission to the decision of the Conference, as it was generally emphasized that after the termination of the Belgian negotiations, a more favourable prospect would be opened to the Poles. Walewski then tried to ascertain whether this would really be the case. He addressed the question to Palmerston who strongly advised the Polish envoy to subscribe to the request of the Belgians. "Les choses parlent par elles-mêmes", he added, "tant que les puissances seront engagées ensemble et conjointement dans les négociations qu'il nous importe beaucoup de terminer, vous concevez que même si nous étions décidés à prendre vos affaires en considération nous ne le pourrions pas." Palmerston said he could not promise anything but did not hesitate to assure the Polish envoy of the genuine sympathy of the Cabinet members and even of those of the Conference. "Et si vous employez l'influence que vous avez sur les Belges pour hâter la conclusion de cette affaire", he continued, "vous agirez (je puis vous donc dire confidentiellement) contre les mesures de la Russie qui tendent au but contraire." In the course of this conversation Palmerston expressed a much more advanced opinion compared to what he had told Granville two weeks earlier, the more remarkable as the Poles had suffered a major

[37] Bülow to Frederick William III, 14 June, to Ancillon, 17 June, Franqué, pp. 121 and 296; Ancillon to Bülow, 7 June, Franqué, p. 289. Cf. above, p. 144.

defeat in the meantime. "Je vous dirai entre nous, que je regarde votre affaire comme gagnée – il me paraît impossible que les Russes puissent, *d'une manière quelconque,* reprendre le Royaume de Pologne. En temps et lieu nous ferons tout ce que nous pourrons." In view of this statement Walewski wanted of course to reopen the question of a consul in Warsaw, but Palmerston disagreed, though he at the same time requested news from Poland as often as possible. In consequence of this interview Walewski promised the Belgians to send a person of distinction to Brussels to plead for the acceptance of new propositions.[38]

Walewski also heard from Talleyrand that England, though prepared to protest in Vienna in Dwernicki's case, refused to do so jointly with France; yet he found "que les Anglais commencent à montrer un intérêt beaucoup plus vif à la Pologne et que des personnes qui étaient tout à fait indifférentes ne le sont plus." The same opinion the French Ambassador conveyed in his dispatch to Paris. Talleyrand wanted Austria to join in the offer of mediation at St. Petersburg and believed that he had created a successful opening through his interview with Esterhazy. Lastly, both he and Walewski observed a marked change in the attitude of the Austrians in London, from Esterhazy and Wessenberg down to the secretaries, towards the Polish envoys.[39]

The news concerning the English attitude was welcomed in Paris, where a speedy termination of Belgian affairs was the order of the day. When the Eighteen Articles were being negotiated Sebastiani, unaware of their tendency hinted that in spite of all its inconveniences the French Cabinet would consider even the partition of Belgium. In any case, the French Foreign Secretary noted that all reports, especially those from St. Petersburg, made the French Government more anxious than ever to see Belgian

[38] Walewski to the Polish Government, 17 June, B.P.

[39] Walewski to the Polish Government, 17 June, to the Paris envoys, 20 June: "Le premier secrétaire de l'Ambassade de l'Autriche vient de passer chez moi pour me féliciter sur la mort de Diebitsch." The Austrians "sont tenus d'une amabilité extraordinaire et se prononcent assez honteusement contre les Russes et se réjouissent ostensiblement de nos succès", to the Polish Government, 24 June: "Il est impossible de témoigner plus de provenance, d'amabilité et d'intérêt que ces Messieurs n'en témoignent, et ils parlent sans détour du désir qu'ils ont de voir nos armes triomfans, ainsi que de la sympathie générale à Vienne." B.P.Walewski was not inclined to overestimate those gestures; it must be observed, however, that they described at least feeling in London. For the English opinion, see Talleyrand to Sebastiani, 15 June, A.é. Angleterre.

affairs terminated, "par un arrangement prompt et pacifique, et s'il est possible avec consentement général, mais *surtout d'accord avec l'Angleterre*". The importance of this condition was made clear by the enclosure to Sebastiani's dispatch, an extract of the "Messager des Chambres" which ran as follows: "On assure qu'un courrier arrivé de Pétersbourg porte la réponse à la démarche faite par le Gouvernement Français pour arrêter l'effusion du sang en Pologne. Tout fait espérer qu'une négociation si importante amènera des résultats, dont la politique et l'humanité auront également à s'applaudir." Sebastiani received the Polish envoys; he regretted that the Polish corps sent to Lithuania was not marching upon Wilno as the Lithuanian insurrection was so important. But Poland was about to become a factor in European diplomacy. The month of July would be decisive.[40]

III

New negotiations in London regarding the affairs of the Low Countries were welcomed neither by Brussels nor by The Hague nor by the Eastern representatives in the Conference. The Belgians, however, at least their delegates in London, could see concessions forthcoming not only in Protocol No. 23, but also in the consideration given to the extraordinary discovery of the Belgian delegate Nothomb, according to whom the "Bases de séparation" accorded the ex-Prussian territory in Limburg to Belgium. As for Prussia, not only Bülow, but also Werther in Paris, surveyed at great length in conversation with the French Minister the possibility of ceding Luxemburg. Eventually, Lieven and Matuszewic yielded to Palmerston's representations of the dangers which would result from the fall of Périer's administration. They accordingly agreed not to send the Belgian delegates away and to try to work out a formula of conditional acceptance by Leopold. They tried to excuse their action by referring to the impossibility of conducting successfully an inevitable general war, with Poland and other events paralysing Russia, Austria and Prussia, thus anticipating their new instructions which forbade them to agree to Prince Leopold's candidature unless the King of the Netherlands should have done so beforehand. They

40 Sebastiani to Talleyrand, 25, 27 June, A.é. Angleterre; The Paris envoys to the Polish Government, 24 June, B.P.

impressed, however, upon Palmerston that not only Belgium but also the King of the Netherlands could reopen hostilities and this would bring about equally disastrous results.[41]

On 22nd June the Dutch Plenipotentiaries addressed a note to the Conference in which they referred to the possibility of coercive measures expressed by the Conference in some of the Protocols, and informed them that the abstention of the Conference from any action left the King with no choice but the employment of means of his own.[42] In spite of new negotiations having been opened under the auspices of warlike feeling in France and Belgium, the Dutch did not expect concessions at their expense; the attitude of the Belgians challenged the Conference and deeply offended the King of the Netherlands who had subscribed to the "Bases de séparation" several months earlier. To their regret the Dutch plenipotentiaries learned that the representatives of the Powers most friendly to their interests took it upon themselves to mediate between the delegates of the Provisional Government, Prince Leopold and the Conference. Granville, they heard, brought from Paris "the most anxious wish of Werther (Prussian *Chargé d'affaires,* generally considered at the time as the successor of Bernstorff) for the Belgian affairs to be arranged, as Prussia could not, because of the Polish war, risk hostilities elsewhere; and Palmerston almost publicly spread this news". The King's speech at the opening of Parliament was "the best that could be expected", wrote Falck, "as it did mention the unity of the five Powers, and yet the radicals will be dissatisfied with the weak expressions employed with regard to Poland." The antipathy towards Russia was growing daily, and Lieven became more cautious than ever. Palmerston himself explained to Grey that he had made Bülow and his colleagues more "manageable" by declaring the necessity of a bilateral agreement with France, should the others fail to agree in time.[43]

But the other Dutch representative, van Zuylen, observed that new instructions obtained by Bülow and Wessenberg had shown them to have gone beyond their instructions in the Luxemburg

[41] Webster, I, 136; cf. Franqué, p. 126; Werther to Frederick William III, 3 June, Franqué, p. 283; Lieven, Matuszewic to Nesselrode, 17 June, Ged. III, 484-6; Nesselrode to Lieven, 4 June, Ged. III, 482.

[42] *Papers Relative to the Affairs of Belgium* (1833), vol. B, p. 48.

[43] Falck to Verstolk, 21 June, Falck Papers, R.A.; Palmerston to Grey, undated but written between the 16th and 23rd, possibly on the 22nd (cf. below, p. 160), Howick MSS.

negotiation. That explained, he thought, why, when subsequently the Belgians went still further in their demands, Bülow and Wessenberg, in agreement with Lieven, who was mostly absent from the meetings of the Conference, had declared to Palmerston and to Leopold, "qu'ils ne se mêleraient plus de cette affaire et qu'ils laissaient le Prince dans toute sa liberté d'agir, sauf à l'Angleterre et à la France de le reconnaître à elles seules, si elles le jugeaient à propos. Celui-ci, étrangé [sic] de cette résolution, s'est empressé de donner grand tort aux Belges." [44]

Bülow thought that his declaration produced a favourable effect; the English had begun to take the Dutch objections into consideration. He moreover believed that the wording of the new propositions, soon to be known as the XVIII Articles, allowed of a "favourable" interpretation, i.e. the breaking-off of the negotiation if the case for it, a better international situation, should present itself. A day later Matuszewic confidentially communicated the Articles to the Dutch plenipotentiaries, begging them to bear in mind that consideration given to this document did not imply its later acceptance; but already on the 26th Prince Leopold received the Belgian Delegates in an official audience, accepting the crown of Belgium, subject to the Belgians agreeing to the XVIII Articles.[45]

The Prussian Minister described the inside story of the Conference. In the course of the renewed negotiations an informal meeting was held at Talleyrand's, with Falck and Van Zuylen present but without Lieven, at which Palmerston represented the necessity of giving up the "Bases de séparation", which the Belgians would not accept, since otherwise a very serious development was inevitable. After the meeting which lasted until three o'clock in the morning Palmerston took him aside to pass on to him some confidential information. He told Bülow that the pressure which the Eastern Powers wanted to bring to bear on the Belgians was greatly disapproved of by the whole Cabinet, and that he was its only member who wished to preserve the unity of the five Powers, whilst all the other Ministers would welcome an arrangement in a liberal spirit with France alone. It had seemed to Bülow that Palmerston accepted his counter-arguments, such as sovereign rights of William I, etc. But a few hours later, on the 23rd very early

[44] Van Zuylen to Verstolk, 21 June, R.A.

[45] Bülow to Waldburg-Truchsess, 27 June, Franqué, p. 126; Falck, van Zuylen to Verstolk, 25 June, R.A.

in the morning, the Prussian Minister received a letter from Palmerston in which he urged making it possible for Leopold to accept the throne of Belgium as the only means of avoiding Belgium becoming a part of France. The argument was supported by expressions of general principle, such as that governments were unable to suppress national feeling, the result being that Napoleon had failed wherever he had to do with nations, "que Charles a perdu la France, Pierre le Brésil, Guillaume la Belgique et Nicklas [sic] la Pologne." It is all right for Russia, wrote Palmerston, to wish to have things embroiled in the West to ensure that no one is able to mix in those in the East; but the Governments of France and England absolutely need peace, and one cannot sacrifice this out of deference to the King of the Netherlands.[46]

Deference was, however, shown to other Powers. Accordingly, the XVIII Articles had not been sent to Brussels by Leopold himself as Palmerston first suggested, since making the Prince the bearer of a proposal made by the Conference would imply the recognition of his acceptance, a course impossible for the Eastern plenipotentiaries; the document was appended to the Conference's letter to Lebeau, the Belgian Minister for Foreign Affairs. The only gesture towards William I was the mission of Wessenberg to The Hague.[47]

More intimate was Palmerston in his letter to Bagot. "The state of France and of other countries makes it absolutely necessary to bring the Belgian affairs to a close." Were the Belgians left to themselves, they would have asked France for some sort of reunion, and no French Government could possibly refuse it. Now, "if the French Government had accepted Belgium we must either have submitted to that... or we must all have gone to war to prevent it; and who would have been ready to prevent that war, not Russia, certainly, for she is too far and too much occupied, not Prussia willingly, for she would have had more to risk than to gain by the jeopardy of her Rhenish provinces; as to us, we might have sent our fleets, but could we have sent an army, sufficient to keep the French out of Belgium?" This being impossible, the "Bases de séparation" had to be abandoned and the XVIII Articles drawn

[46] Bülow to Frederick William III, 25 June, Palmerston to Bülow, 23 June, Franqué, pp. 301, 308.

[47] Palmerston to Grey, two letters undated, but written on 25 and 26 June, Howick MSS.

up, to enable the Prince of Saxe-Coburg to accept the Belgian crown immediately.[48]

Bülow, however, apparently was not sufficiently aware of Anglo-French difficulties. As the final argument, which made him take upon himself the bulk of the preparation of the XVIII Articles, he declared that non-participation of the Eastern Powers in those proceedings would ensure the recognition of Poland by France and England. And even now, he believed that only the personal sympathy of Grey for Mme Lieven made Grey suppress his feeling with regard to the Poles.[49]

Shortly after learning of the XVIII Articles the Dutch plenipotentiaries protested to the British Ministers against what they considered to be an act offensive to the King since he had subscribed to the "Bases de Séparation" several months earlier. They observed an apologetic attitude on the part of both Grey and Palmerston when questioned as to the contents of the new propositions. No easy task accordingly awaited Baron Wessenberg who was delegated by the Conference to bring them officially to the knowledge of the King. The Netherlands Minister for Foreign Affairs did not, as on previous occasions, conceal from the British Ambassador that not only were these propositions, as reported by Falck and Van Zuylen, unacceptable, but that the King, on learning of their contents, would, "immediately carry into effect the project which he had been some time under the necessity of contemplating", namely, to order his troops to march into Belgium and his ships to block the Scheldt. Arguments concerning the peace of Europe, which would have been threatened by coercive measures with regard to Belgium, had an opposite effect. "The Five Powers were grievously mistaken", Verstolk told Bagot, "if they supposed that the peace of Europe was not quite as much in the King's hands as in those of the Congress of Belgium." Bagot took a rather serious view of this conversation. Wessenberg would be received and the propositions heard, but if they contained anything calling for further sacrifices of purely Dutch interests, "they will be flatly refused and hostilities probably commenced." The King also contemplated passing on some information to the States General, "which would be very embarrassing to other States".

[48] Palmerston to Bagot, 27 June, Bagot MSS.
[49] Bülow to Frederick William III, 28 June, Franqué, p. 132.

"I have, as you know, had great doubts of the sincerity of the threats held out to me to this effect", concluded the British Ambassador in his long account, "but though unofficially, Mon. Verstolk here spoke the King's sentiments and what he said to me this morning is seriously meant." [50]

Two days later Bagot softened his report, saying that Falck and Van Zuylen probably exaggerated in their accounts of the new propositions and that Wessenberg's explanations might be viewed in a better spirit. The Austrian delegate himself, almost immediately after his arrival, found it impossible to share this hope. In his first account he described the feeling in The Hague as very warlike. He found both people and the Government unaware of the real state of things. Besides, he observed "la fausse conviction que le sort de la Hollande sera décidé sur les bords de la Vistule" – that after the suppression of the Polish revolution the Emperor of Russia would put the whole weight of such forces as were left in favour of Holland.[51]

At the same time diplomatic reports from Prussia, always closest to The Hague, said that the King of the Netherlands was unlikely to accept the XVIII Articles or to consider a pecuniary indemnity for Luxemburg. Diplomacy, however, was then still concerned with the possibility of a Belgian refusal only.[52]

IV

The acceptance of the XVIII Articles by the Belgians was no simple matter. According to his promise Walewski sent Count Załuski, related to a Belgian family, d'Aerschot, to Brussels to use his influence for the purpose.[53]

Before the reports from Brussels warranted any hope that the Belgian Congress would agree to the new proposals, preliminary steps in the Polish affairs seemed to have been taking place. The

[50] Van Zuylen to Verstolk, 27 June, Falck to Verstolk, 28 June, R.A.; Bagot to Palmerston, 28 June, Bagot MSS.

[51] Bagot to Palmerston, 1, 2 July, Bagot MSS; Wessenberg to Metternich, 3 July, Ged. III, 201.

[52] Chad to Palmerston, 5 July, F.O. Prussia; Flahaut to Sebastiani, 9 July, A.é. Prusse; cf. Falck to Verstolk, 27 June, R.A.

[53] An account of Załuski's activity in Belgium is given in F. Perelman-Liwer, *La Belgique et la révolution polonaise de 1830*, 1948, pp. 31-41.

first encouraging news, concerning the attitude of France reached the Polish agents in Paris through Berlin and London. However, Sebastiani told Kniaziewicz that the Poles must help French policy by a repeal of the deposition of Nicholas. Thereafter Périer received the Polish envoy, who thanked him for what the French Government was doing for the Poles, for semi-official newspapers were always hinting that negotiations concerning Poland were being carried on. Périer gave no details, except "qu'il ne s'agissait pas de soumission." Kniaziewicz then said that money and arms, if supplied to the Poles, might make diplomatic action superfluous. "Cela s'arrangera", replied the French premier. Reports from Belgium improved. From London Walewski wrote that most of the Tories had lost their prejudice against the Polish revolution. He spoke to Aberdeen and his impression was that no opposition was to be expected from that quarter.[54]

Before this report reached Paris, Plater and Kniaziewicz addressed another note to Sebastiani, in which they referred to his previous statements. On the following day, 8th July, the Minister received them. He proudly informed the Poles of his note sent to London, inviting the British Government to joint action at St. Petersburg. On this occasion Sebastiani had much to say. He spoke of Mortemart's action at St. Petersburg and of Flahaut's at Berlin. Disarmament, protests concerning the corps of Dwernicki, this all revolved around the new situation. That Palmerston did not want negotiations until November seemed an obstacle likely to be removed. The conclusion was impressive: "Mon but est d'informer aussi promptement que possible le Prince Adam [Czartoryski] et le Gouvernement. Je veux leur donner de la force – qu'ils tiennent encore deux mois – c'est nécessaire pour les négociations." [55]

When the French courier had left for London, Sebastiani saw the British *Chargé d'affaires* (Granville being away in England) and read to him his instructions for Talleyrand. Mr. Hamilton replied that there was enough sympathy for the Polish cause in England to expect concurrence in the French proposals, were there any hope that they would be well received in St. Petersburg. If Nicholas expected to subdue the Poles during this campaign, he would not listen to any interference; it would be a different matter

[54] Walewski to the Paris envoys, 28 June, 5 July; the Paris envoys to the Polish Government, 1 July; Załuski to the Paris envoys, 3, 5 July. B.P.
[55] The Paris envoys to the Polish Government, 7, 8 July, B.P.

should the campaign end without any positive result. The same answer Mr. Hamilton returned to Louis Philippe, who spoke to him on the subject. "If you reject the proposal", said the King, "it will be for you who reject it to justify such rejection to your Parliament and to your country; we at least shall be able to show that we have done what we could to avert from Poland the destruction impending over that unfortunate country." [56]

Again, the observer from the other side saw much more in the French *démarche*. Baron Fagel, the Netherlands Minister, believed that the certainty of its having been made very considerably influenced the course of the elections which seemed to assure a majority for the Government. He had been assured that the Government resolved to inform London, Berlin and St. Petersburg of its determination to put an end to the bloodshed between the *two nations* on the Vistula; he believed that this step had been agreed upon between England and France and moreover he considered it of importance that Kniaziewicz had paid a visit at the Palais Royal in uniform. The Poles themselves heard from Sebastiani that he expected the English reply within 5 or 6 days but thought it unnecessary to wait with the dispatch to Warsaw until then.[57]

The first report from Talleyrand concerning the "Polish" note was to a certain degree encouraging. He spoke about it to Grey, Palmerston and other ministers. "Je me suis servi utilement de tous les raisonnements contenus dans votre dépêche du 7me; ils m'ont paru faire impression sur les Ministres." Nevertheless, the matter was to be considered by a Cabinet meeting. Talleyrand would do his best, not without some anxiety, before the meeting of the 13th. On the 13th it appeared that the Cabinet dealt exclusively with coronation arrangements. Sebastiani, however, was still hopeful about the English reply. The Polish envoys considered the attitude

[56] Hamilton to Palmerston, 8 July, F.O.France.

[57] Fagel to Verstolk, 8 July: "L'on m'assure que le Ministère, à la demande du Roi, a résolu de faire connaître à Londres, Berlin et St. Pétersbourg, sa détermination de voir une fin à l'effusion de sang entre les *deux nations*, Polonaise et Russe. MM. de Mortemart et de Flahaut recevront incessamment des instructions dans ce sens. Un fait certain, en attendant le résultat de cette résolution, est, que le Général Kniaziewicz a été admis ce matin même à faire sa cour au Roi et à la famille Royale, et qu'il s'est rendu au Palais Royal en uniforme." R.A.; The Paris envoys to the Polish Government, 8 July: "Le Général Kniaziewicz a été parfaitement accueilli chez le Duc d'Orléans." This was probably the only member of the Royal family he saw, and the Poles paid less attention to the fact than did the Dutch Minister. B.P.

of Prussia changed, when they learnt that Orlov, on a special mission to the Prussian capital, had to leave Berlin within a few hours without seeing the King; at the same time Werther and Pozzo heard some harsh words from the French Minister. The Russian Ambassador had to swallow the reproof that England and France could not remain indifferent to the slaughter of Poland any longer and that they would recognize the independence of Poland. Moreover, the XVIII Articles were accepted by the Belgians which Sebastiani called "une bataille gagnée". The Poles knew that Lafayette had spoken to Périer and Soult on the subject of sending a naval squadron to the Baltic. Mortemart was informed that the acceptance of the XVIII Articles by the Belgians enabled the French Government to devote their efforts to other questions. "Ceux qui se rapportent à la Pologne," commented Sebastiani, "auront, vous le comprendrez facilement, une partie principale dans notre sollicitude." [58]

Meanwhile, the indispensable support of England did not appear to be forthcoming. Grey, who openly declared the absolute necessity for European peace of Périer's remaining in power, was reminded that Périer's government would not survive the fall of Poland, but regretted the lack of a suitable pretext to interfere in Polish affairs. The only result of two Cabinet meetings was the request made to the French Ambassador for an official note on the subject and for the British Government to be confidentially informed what France proposed to do for the Poles. On the 15th Palmerston saw Walewski, and this interview clearly showed that as far as Palmerston was concerned only a negative answer could be expected. He promised to send to Prussia a protest against the infringement of neutrality, similar to that already sent to Vienna with regard to Dwernicki's corps; but he also most clearly expressed his positive conviction that unless the first campaign should show the inability of Russia to terminate the contest by her own means, Britain could have no hope of a positive reply were she to offer to mediate. Yet Talleyrand, on learning of this interview, did not lose his confidence: "Je n'aurai pas de refus", he told Walewski.[59]

[58] Talleyrand to Sebastiani, 11, 13 July, A.é. Angleterre; Sebastiani to Mortemart, 14 July, A.é. Russie; The Paris envoys to Walewski, 11 July, to Załuski, 11 July, B.P.

[59] Walewski to the Paris envoys, 13, 14, 15 July, to the Polish Government, 18 July, B.P.

He had at least one convinced supporter in the British Cabinet. Lord Holland saw in the termination of the Belgian proceedings a prospect for an improvement of relations between France and England. He was anxious to see an agreement on common action with regard to Poland and considered it equally important as a means of strengthening the position of Périer in France and of Leopold in Belgium.[60]

Little other evidence of a favourable reception of the French proposal seems to be traceable. Sebastiani pressed the British Ambassador, stressing the weight of representations if made at St. Petersburg conjointly; some effect, he believed, had already been produced by the French protests alone, and he added that France would continue along this line. He even hinted at the desirability of a maritime action, an argument, it will be seen, successfully used by Flahaut in Berlin. Lastly, the French Minister referred to parliamentary difficulties. Of this there was enough evidence, and the Polish envoys observed that as the results in Belgium were not satisfactory, the Ministry needed support by the development of Polish affairs; nevertheless they believed that without the concurrence of London nothing would be done by France.[61]

On the 18th Sebastiani began his conference with the British Ambassador "by complaining of the disinclination of H. M. Government to join that of France in their representations in favour of the Poles." Any delay would be regrettably fatal. Granville was not impressed: if matters were so bad, he said, Russia would not suffer intervention. The French Minister then turned the conversation to Belgian affairs, arguing for a settlement concerning the demolition of the frontier fortresses. The reply was that there was no reason "that any sacrifices should be made by England because the election of Prince Leopold has been favoured by France." "We shall be attacked in the Chamber", observed Sebastiani. The same complaint Granville heard from Périer while enquiring about the contents of Louis Philippe's speech to be made at the opening of Parliament. The French Premier, with more than usual reserve, answered that nothing had as yet been decided upon. He was disappointed that the proposals concerning Poland were so coldly received and no

[60] Holland to Grey, 10 July, Howick MSS, to Granville, 13 July, Granv MSS.

[61] Granville to Palmerston, 15 July, F.O.; The Paris envoys to Walewski, 17 July: "On nous fait entendre ici que l'Angleterre, contente d'avoir atteint son but à Bruxelles – jouera la France." B.P.

answer given. "He said the French Government must meet the wishes of the Chamber and of the Country, and that, although England might decline to join in with France, the French Government must renew its representations to the Emperor of Russia." To this the Ambassador merely replied that as no excitement with regard to Poland existed in England, the Government was perhaps better placed to judge the advisability of an interference at present. The Polish observers in Paris and London were on the point of admitting the English refusal and the powerlessness of France.[62]

Lord Holland seems to have been the only Minister who warmly advocated yielding to the French request. He believed that representations made jointly by England and France would carry enough weight to make Nicholas consider them very seriously. As regards the danger to peace, he thought that the influence of England would contribute to moderation in the common approach to Russia, while France, left to herself, might be pushed too far by popular feeling, especially as the Government's position was weakened by the election of Prince Leopold in Belgium. On the other hand, Austria and Prussia would not concur in a war against France, supported by England, and Russia would not wage war on her own. Further, Lord Holland believed that Austria and even Prussia would not really be opposed to the entire independence of part of Poland, which he considered to be the potential result of the proposed proceedings. Lastly, he introduced a new view of the problem. Because of French interest in Poland that country was more or less regarded as the bulwark of France. When it should be known that France had shown more interest in Poland than England did, public opinion would compel the Government to join in the French intervention; but the initiative having been taken by France, "any Power between Germany and Russia which may be revived

[62] Granville to Palmerston, 18 July, F.O. France; The Paris envoys to Walewski, 19 July: "On commence à dire que la mention polonaise dans le discours de la couronne pourra être faible... Sebastiani est redevenu froid et silencieux. On parle de la sortie du Ministère. M. Pozzo soutient que nous soyons encore une fois dupe par la France, et que son intervention ne sera qu'une bravade." Walewski to the Polish Government: "...il n'est que trop vrai que Ld Grey continue à se montrer assez froid pour notre cause." 18 July. B.P.; Pozzo reported on the 14th: "Dans un but électoral le gouvernement français crut devoir répandre la nouvelle que l'intervention française en faveur des Polonais était favorablement accueillie par le cabinet russe." (to Nesselrode, 14 July, Martens, *Recueil*, XV, 135). The course of the negotiations shows, however, that the electoral motive, even if it was the only one, became a very pressing one indeed.

in consequence of such interference will form more emphatically part of the French rather than English connection in Europe." [63]

On 18th July Sebastiani instructed Talleyrand to present to the British Government an official note concerning Poland, and, if any sign of a favourable reception were shown, to propose joint efforts for a suspension of hostilities in Poland. The Ambassador replied that he had done so, had spoken to Palmerston, Grey and other Ministers, but obtained no positive result: the answer was that the appropriate moment had not yet arrived. By the same dispatch he sent word that the news of the rejection of the XVIII Articles by the King of the Netherlands had arrived in London.[64]

Yet, the other side of the Conference was by no means easy with regard to the attitude of the Western Powers in Polish affairs. Firstly, one thing was becoming clear to the Russians: the growing antipathy of the public towards their country; the Russians were aware that the British Government had seriously to consider whether they were strong enough to challenge public opinion by doing nothing about Poland. Matuszewic, the other Russian representative had been several times attacked in the newspapers as a renegade Pole. The Lievens did not succeed in excluding the subject of Poland from the King's speech to Parliament and realized that Grey was concerned in it through his well-known pro-Polish pronouncements in the past; Mme Lieven was inclined to advise the Grand Duchess Helen to renounce her visit to London.[65]

In France the publicity given to the article in the "Journal de St. Pétersbourg" provided a powerful stimulus to anti-Russian feeling. But since the article supplied further evidence that Poland, though her revolution was becoming more than ever the bulwark of France, this was one more reason for expecting that the initiative in the defence of Poland would be taken by the Paris Government. Matuszewic, reflecting on this article, observed that it contained no evidence of any definite hostile plans on the part of Russia with regard to France; but he connected the subject with another

[63] Holland to Granville, 19 July, Granv MSS.

[64] Sebastiani to Talleyrand, 18 July, Talleyrand to Sebastiani, 20 July, A.é. Angleterre.

[65] Mme Nesselrode to her husband, 5, 17, 20 July. The third letter reads: "Il faut être à l'étranger pour se figurer à quel point nous sommes en ce moment haïs et poursuivis... Que nous ayons en Pologne des succès ou des revers, cela ne fera qu'accroître l'animosité qui règne", Nesselrode, *Memoirs*, VII, 188, 202, 203.

report which bore reference to the policy of the British Government. After referring to the article he noted in the same paragraph his meeting with two very important Tories, who had declared to him that they knew of the English Government's intention to get in touch with that of France in order to interfere in the affairs of Poland. Talleyrand led the intrigue, they told him, and they smiled at his incredulity.[66]

On Lieven's orders Palmerston was sounded by Matuszewic a few days later on the contents of the above information. Matuszewic found the answer very satisfactory. The Foreign Secretary not only refrained from any comment on Mortemart's set-back at St. Petersburg early in June, but declared that England had neither the right nor the means to intervene in favour of the Poles. At the same time, however, Palmerston added that he did not believe in the ability of Russia to reconquer the Kingdom of Poland or to keep Lithuania. Matuszewic made an effort to show the contrary, and seemed confident of his success; he seized the occasion to point out the impropriety of an Englishman's relations with the Polish agents. In the same letter Matuszewic reported the departure of Biré and the negotiations of the Poles concerning loans, the purchase of arms and their transport to Poland. On the same day Matuszewic rendered an account of his interview with Palmerston to Mme Lieven; Palmerston's opinion was unequivocally put in a single sentence: "que l'Angleterre n'avait ni le droit ni les moyens d'intervenir en faveur des Polonais, mais qu'il [Palmerston] regardait la Pologne et la Lithuanie comme perdues pour nous." [67]

The possible intervention in Poland continued to be the current news throughout the first half of July. Polish reports from Brussels quoted Lord Erskine's opinion, made known during his passage to his post in Munich, that Grey would intervene as soon as the Belgian affair had come to a close. The Netherlands Ambassador in London was certain of negotiations already taking place. He

[66] Cf. above, p. 143; Dutkiewicz, pp. 109-110. Matuszewic to Lieven, undated, written shortly before June 25: "Deux Torys *très marquans* m'ont entrepris hier sur les affaires de Pologne et déclaré qu'il était parvenu à leur connaissance que le cabinet anglais cherchait à négocier avec la France les moyens de se mêler de nos affaires et de nous imposer un arrangement. Je me suis contenté de leur répondre qu'ils devaient être en erreur, etc... Cependant on a souri de mon incrédulité et on m'a fait entendre que Talleyrand est l'intermédiaire de cette intrigue..." Lie MSS.

[67] Matuszewic to Lieven, 1 July, to Mme Lieven, 1 July, Lie MSS. For Mortemart's set-back, cf. below, p. 187.

considered Talleyrand's insistence more pronounced in Polish affairs than in the question of the demolition of the Belgian fortresses. The purpose of his efforts seemed to be either the sending of a naval squadron to the Baltic or an approach to the Emperor Nicholas with a view to placing a Prussian Prince on the restored throne in Warsaw. Additional evidence of English concurrence with these plans, Falck thought, was furnished by the changed attitude of "The Times"; this journal, in which Brougham had become most influential, became pro-Belgian after the acceptance of the XVIII Articles by the Brussels authorities, and this acceptance was the condition *sine qua non* for the Government to turn its attention to Poland. Yet, "the Russian Embassy which is informed about it", wrote the Netherlands Ambassador, "is much more confident about the feelings of Lord Grey and his colleagues, than I can afford to be, knowing the political opinions of the British Cabinet." [68]

A similar situation existed in Paris, where, however, the general opinion was that the Government must make a statement with regard to the intervention, even at the risk of war. The Netherlands Minister believed that the King's speech at the opening of Parliament would be enough – for the time being. In the meantime, the newspapers continually reported the departure of the Russian Ambassador, who, however, remained in Paris.[69]

By the 15th, the choice of the British Government seemed to have been made. Palmerston assured Princess Lieven that, even if intervention were possible, it would not be offered at a critical moment. A similar declaration was confidentially made to Lieven, who, however, on learning of Talleyrand's note, hastened to acknowledge the good faith of the British Government by expressing the hope that the French proposals would be formally refused.[70]

On the 19th the Netherlands Ambassador reported that no more doubts were entertained as to the refusal of the British Government. Talleyrand, wrote Falck, had expected a more favourable reception for his project, in return for the French acquiescence in Prince Leopold's election in Belgium. The French Ambassador

[68] White (British Consul in Antwerp) to Załuski, 14 July, B.P.; Falck to Verstolk, 12, 14 July, Ged., IV, 529-531.

[69] Fagel to Verstolk, 18 July, R.A.

[70] Mme Lieven to Nesselrode, 14/15 July. The draft of the "Lettre confidentielle" is appended to Matuszewic's letter to Lieven of 21 July, Lie MSS.

referred again to the necessity of Périer's Cabinet having to reckon with the opinions of the *parti du mouvement;* but Grey was said to have answered that there was no difference in the position of the British Government in this regard, and they felt no such necessity.[71]

The preparation of an answer to Talleyrand's note engaged the attention of the Cabinet during two days. On the 20th Palmerston sent his draft for the approval of the Cabinet. Grey answered the same day that, although the Ministers generally agreed with it, a special meeting would be held to consider a matter of such importance. Lord Holland as usual dissented, demanding that, if a refusal was unavoidable, the note should be rendered "more palatable to France and uncomplimentary to Russia"; in its conclusion it should be said that England would gladly seize any opportunity, in concert with France, to stop the bloodshed in Poland short of one incompatible with the friendly relations with Russia "and the recent cooperation of that Power in adjusting the differences between Holland and Belgium." Here, however, Palmerston's view varied from that of the other Ministers. He objected to the passage mentioning the loyal attitude of Russia as a motive to refrain from intervention. Should, within 6 or 8 months, the intervention become necessary, he thought, Russia would then surely refer to it.[72]

The majority view, however, was represented in the official reply, dated 22nd July. It repeated to a certain extent what had already passed between Talleyrand and Palmerston, and what had been included in the latter's instructions sent to St. Petersburg. It was emphasized that, as far as the Treaty of Vienna was concerned, "His Majesty could not consent to see Poland deprived of the advantages of that arrangement," and H.M. Government had already made appropriate representations to the Imperial Cabinet to that end. But at the same time H.M. Government feared that any offer of mediation would be refused, and would only leave to the two governments the embarrassing alternatives of either acquiescing in a determined rejection of their proposal or of taking measures to enforce it by means of a more direct and effectual interference. The British Government certainly was not prepared to adopt the latter course, "warranted neither by the effect of the contest in

[71] Falck to Verstolk, 19 July, R.A.

[72] Palmerston to Grey, 20, 21 July; Holland to Grey, 20 July, Howick MSS; Grey to Palmerston, 20 July, Palm MSS.

other countries, nor by the attitude of Russia towards England; on the contrary, especially in the late difficult negotiations concerning the separation of Holland and Belgium she acted with perfect fairness and co-operation." "Under these circumstances", proceeds the document immediately, "His Majesty, deeply lamenting the calamities of a disastrous and desolating contest, does not think the time has yet arrived when he could be justified in adopting a proceeding which, however conciliatory in form, could not fail to alarm an independent Power, naturally jealous of its right and sensibly alive to everything which might appear to effect its national honour." However, though the French proposal was declined, "the King can never look with indifference on such a state of things which now exists in Poland and will avail himself of any favourable opportunity which the friendly relations subsisting between the Governments of Great Britain and Russia may afford", to help the restoration of peace in the countries concerned.[73]

The note was satisfactory to Russia in that it refused to take part in an intervention; considering the state of the war in Poland as well as for speculations on the subject in diplomatic circles, this was in itself a victory for St. Petersburg. At the same time, however, it constituted an almost public declaration of the British Government's interest in Polish affairs in consequence of their having been the subject of an international agreement, a statement which met with a cold reception when made in a confidential way by Heytesbury in April.

If, under the circumstances, this declaration attracted little attention, the passage referring to the part played by Russia in Belgian affairs could not fail to impress the parties concerned. This point is raised by Talleyrand in his report to Paris. Van Zuylen saw the connection of Belgian and Polish affairs as so obvious as to believe the existence of an unwritten pact of mutual non-interference between London and St. Petersburg. And the French contemporary showed offence, commenting upon the paper as "le refus dont la forme polie ne couvrait qu'imparfaitement l'insolence. Ici encore M. de Talleyrand venait d'être joué." Talleyrand indeed, as it has been shown above, expected no refusal. On the English side the paper was the subject of regret to Lord Holland.[74]

[73] B.F.S.P., XXXVII, 1413–1415.

[74] Talleyrand to Sebastiani, 23 July, A.é. Angleterre; Van Zuylen to Van Maanen, 23 July: "The failure of France to move England to a *démarche* in

V

The Belgian delegates in London were not unmindful of the influence of Załuski in Brussels at the time when the XVIII Articles were under consideration. Shortly before their return and before the final answer of the British Government to Talleyrand's note on mediation was known, Lebeau and de Merode went to see Palmerston. They pointed out the part played by Polish affairs in the discussions of the Congress. The Foreign Secretary, however, "se borna à répondre que c'était une question très compliquée." Grey also spoke in vague terms.[75]

When the British refusal became known the Polish envoys called on Sebastiani. The Minister tried to encourage them. The news from Poland was not bad and would no doubt influence Flahaut's activity in Berlin. Moreover, the reference to Poland in the King's speech would play its part in Berlin, in Vienna, and, above all, in St. Petersburg. In London Walewski met with an excellent reception from Palmerston, who explained to him the reason for refusing the French offer: it had no chance of being favourably received by the Imperial Government. But the British Government was following the development of events in Poland with the greatest attention, and as soon as an occasion should present itself it would gladly make use of it to act in favour of the Poles. Walewski then immediately referred to the neutrality of the neighbouring states, especially of Prussia. Palmerston replied that the conduct of Prussia had already been criticized and that the British Minister in Berlin had been instructed to approach the Government on the subject. The British Government maintained that Prussia not being at war with Poland, any goods which British subjects might wish to send to Poland, even arms and munitions, must be allowed to pass through Prussian territory.[76]

---

favour of the Poles is the result of a contractus nec nominatus of facio ut facias between London and St. Petersburg: I will not meddle in your Polish affairs if you help me in my Belgian projects." Ged., IV, 537; L. Blanc, *L'histoire des Dix Ans*, 1844, II, 451; Holland to Granville, 28 July: "I was in hope since we did not join in the offer of mediation that our manner of declining it would be almost as agreeable to the French as our agreement would have been." Granville MSS.

[75] Załuski to the Paris envoys, 22 July, B.P.

[76] The Paris envoys to the Polish Government, 26 July, Walewski to the Polish Government, 29 July: "... si la Prusse n'intervenoit positivement en faisant la guerre aux Polonais ... nous entendions que toute espèce de marchandise

The situation on the spot, however, hardly, corroborated the above reasoning. It has already been emphasized that Prussia experienced the greatest difficulties on account of the revolution in Poland; from every aspect she was most interested in its termination. The Prussian Government accordingly tried every means at its disposal to bring about a speedy pacification of Poland. Prussian attempts at mediation have already been noted. Interception of correspondence, obstructions for travellers, and, latterly, direct assistance afforded to the Russian forces in the form of supplies had occurred. These circumstances were observed by the French and British representatives in Berlin; Chad accumulated a good deal of evidence from the reports of the British Consuls in Memel and Danzig, as well as from other quarters, during the spring months of 1831.[77]

The war in Poland made it necessary for Prussia to keep a large force on her Eastern frontiers. Much in the same way as Metternich, Bernstorff also represented Prussian military preparedness as only a necessary response to the situation in neighbouring countries. When in May the French Government proposed talks on disarmament, the Minister expressed his regret that the state of affairs did not allow of such proceedings. The Prussian Government was most sensitive to the financial pressure created by military expenses and was most anxious to relieve it, but not before Belgium and Poland were entirely settled. It seemed that in Belgium that stage was about to be achieved, but not in Poland where war was still being carried on, much to the damage of Prussian interests. The result of the battle of Ostrolenka was represented as far from decisive. Similar opinions were held out to Chad by Ancillon who moreover expressed the opinion that Russia might decline disarmament talks merely for fear that an assembly for that purpose might take up the subject of Poland. On this occasion Berlin became again uncomfortably aware of the British opposition to a possible Prussian intervention in favour of Russia. An application by Diebitsch for

que nos sujets pourraient avoir intérêt de faire passer en Pologne même avec le but avoué d'aider les Polonais et *sans excepter les fusils et autres munitions de guerre* y passent sans aucun embarras." B.P.

[77] Mortier to Sebastiani, 9 May, A.é. Prusse; Chad to Palmerston, in particular 30 May, 5, 9 June. Consul Brockmann reported from Memel on 21 April that a detachment of 800 Russians marched through that town. F.O.Prussia; Mortier wrote on 30 April that Ancillon showed embarrassment when told that some Russian troops, though disarmed, were allowed to pass through Prussian territory. A.é. Prusse.

permission to make use of the bridge over the Vistula on Prussian territory was promptly rejected.[78]

It was probably the belief that Prussia would welcome an arrangement in Poland which prompted the French Government to send Count Flahaut to Berlin on a special mission. The idea was reflected in his instructions, dated 28th May. As public opinion in Germany was in favour of the Poles, the Prussian Government might perhaps be disposed to offer its good offices in St. Petersburg, which would contribute considerably to a settlement of this embarrassing affair. Moreover, there were really no common or opposed interests between Paris and Berlin; why should not Prussia help France with a problem which was a subject of constant anxiety to the French Government? This point was put in elaborate explanations aimed at showing the alleged absence of opposed interests. It is, however, easier to believe and indeed more likely that the French Minister used delicacy towards the Prussian Government by avoiding stress upon questions that might strike Berlin painfully.[79]

Flahaut arrived in Berlin on 7th June and soon obtained his audience of the King. "Le Roi a parlé alors de la révolution de Pologne avec peine", wrote the French Ambassador. The King said that after the battle of Ostrolenka neither party seemed able to do anything, Diebitsch staying where he was and the Poles engaged in recovering from their great losses. Flahaut then seized the occasion to refer to an intervention. Considerations of humanity alone were no doubt likely to make Fredrick William III attempt to bring an end to this struggle, in which case the King of the French would be only too glad to concur. The King of Prussia seemed to agree, but evidently the subject was painful to him and the Ambassador thought it better to drop it. Ancillon also spoke on Poland regretfully but the idea of mediation was foreign to him: submission or a Russian victory were the only ways open at this stage to terminate this unfortunate affair.[80]

After two weeks in Berlin Flahaut formed his own opinion:

[78] Mortier to Sebastiani, 31 May: "(Polish affairs) ... étaient subordonnées aux chances d'une lutte opiniâtre dont malheureusement la Prusse éprouvait plus que toute autre puissance, de graves et sérieux embarras. Le Comte de Bernstorff est loin de supposer comme décisif l'avantage remporté par le M^{al} Diebitsch." A.é. Prusse; Chad to Palmerston, 24 May, F.O. Prussia, 18 May, Palm MSS; Alopeus to Nesselrode, 16 May, Martens, *Recueil,* VIII, 173.

[79] Sebastiani to Flahaut, 28 May, A.é. Prusse.

[80] Mortier to Sebastiani, 9 June, Flahaut to Sebastiani, 11 June, A.é. Prusse.

the Prussian Government was unable to afford any more direct assistance to Russia, because of the course of public opinion inPrussia itself. It even made some ostensible gestures to furnish the appearance of impartiality. Thus Orlov, who brought to Berlin the plan for the partition of the Kingdom of Poland – in itself evidence of Nicholas' depression – was sent away within a few hours on the ground of his having broken quarantine regulations. News of this and similar steps made its way to Paris and London, though perhaps it did not deceive many; Greville at least had no doubt that help was being given to the Russian army.[81]

On the other hand, there was no question of concurring with any steps which aimed at altering the formal status of the Kingdom of Poland. Berlin was even more sensitive than St. Petersburg to Western interest in the subject and opposed to the idea of foreign interference. The King therefore advised Nicholas to grant to the Poles a status similar to that enjoyed in the Austrian Empire by Czechs and Hungarians. Though unaware of this advice Flahaut tried to exploit the sensitiveness of Berlin in this respect. The article in the "Journal de St. Pétersbourg" served as a welcome pretext and the Ambassador made a point of it. He told Bernstorff that this was a very serious matter: if there was no sufficient apology, France might perhaps recall her Ambassador from St. Petersburg, or even send a fleet to the Baltic. Flahaut was sure that the latter suggestion made a profound impression on the Prussian Minister.[82]

The Ambassador accordingly lost no time in conveying his impressions of Prussia to the knowledge of the British Cabinet, passing over Chad, a Tory, from whom he expected no co-operation. "Our whole conduct towards Russia has done much harm", he wrote to Grey. All the continental Powers were unlikely and unable to risk a war, and would therefore withdraw before threats. Russia was no longer a Power, he thought, and the Prussian Government especially felt very strongly the truth of this for several years to come. [83]

[81] Flahaut to Sebastiani, 23 June, A.é. Prusse; Schiemann, *Geschichte Russlands*, III, 119-122; Chad to Palmerston, 18, 21 June, F.O. Prussia. Chad believed that Orlov came to obtain more facilities for the passage of provisions; *Greville Diary*, 8 and 13 July, 1831, II, pp. 160, 163. Cf. also above, p. 165 and below, p. 178.

[82] Maltitz to Nesselrode, 22 June, Martens, *Recueil*, VIII, 177; Flahaut to Sebastiani, 25 June, A.é. Prusse. For the article in the "Journal de St. Pétersbourg", see above, p. 143.

[83] Flahaut to Grey, 28 June, Howick MSS.

Flahaut himself might not entirely subscribe to the above view, but other reports from Berlin seemed to support the information he had sent to London. From the inactivity of the Russian army Berlin military circles concluded that Paskevitch was unable to attack Warsaw, noted Chad in the beginning of July. This was also a reason for Flahaut to hope that St. Petersburg might now be more disposed towards conciliation, he added. And the new ukase concerning Lithuanian insurgents led the French Ambassador to the same conclusion, which, however, was not shared by his British colleague.[84]

On 14th July Flahaut began his final attack with regard to the Polish policy of Prussia. The Prussian Government, said Bernstorff, was not neutral, only inactive. Russia could have asked for assistance to re-establish Nicholas' authority in Poland, and she did not. Now the re-establishment of the Kingdom of Poland, as demanded by the Poles, would cause endless troubles in Europe, as they would ask for Lithuania, Galicia, Thorn, Danzig etc; even if the Poles enjoyed the sympathy of the peoples, politicians could not listen to the voice of the heart. To all this the Ambassador replied that, on the contrary, sentiment did play a part in French politics at least, and among the German people as well, upon whom assistance given to the Russians created a poor impression. On the other hand, he thought that Nicholas would be glad to have mediation offered by Prussia as she was on quite different terms with St. Petersburg than other Courts. Bernstorff did not share this opinion; he thought moreover that the termination of the contest by a negotiation was no longer possible, in view of the extent of the revolution.[85]

On a subsequent occasion Flahaut went even further. The attitude of Prussia showed more than "inactivity". He said he did not doubt that an armed intervention by Prussia would be counteracted by French military action. Then followed an exchange of views on non-intervention in which Belgium served as an example. Matters need not go that far, he hoped, to which Bernstorff replied that he wished that Russian successes, if necessary at all, could be delayed long enough for news of them not to coincide with the opening of the French Parliament.[86]

[84] Chad to Palmerston, 3, 5 July, F.O.Prussia.
[85] Flahaut to Sebastiani, 14 July, A.é. Prusse.
[86] Flahaut to Sebastiani, 18 July, A.é. Prusse.

The British Envoy remained much more reserved. On the occasion of one of his interviews regarding a British subject who was refused passage to Poland, Ancillon tried to dispel his interest in Poland. Chad's modest comment was: "I do not think that Mr. Ancillon makes a just distinction between the insurrection in Belgium and in Poland, although he does justice to the bravery of the Poles, and I tried in vain to bring him over to my way of thinking upon that subject." Chad had even less success with Bernstorff, who told him that "Lord Grey feels upon the subject with all the warmth of a student." In his turn, Flahaut failed to enlist the co-operation of the British Minister. The Ambassador's account of his interview with Bernstorff was commented on by Chad as "warm but not hostile", and sent to England by a safe channel, because if it came to the knowledge of the Prussian Government, "an appearance of concert in the matter might arise," remarked Chad, "which I wish to avoid as I think the French Government would not be sorry to estrange Prussia from England."[87]

Flahaut believed that his representations had some effect on the Government, but at the same time he reported that assistance given to Russia had been exaggerated; meanwhile, the sanitary cordon had made supplies almost cease. Besides, if France continued to press Berlin for a more pronounced attitude in the Polish question, Prussia might eventually turn to open assistance to Russia. Shortly after, however, he was assured that Paskevitch had been authorized to terminate the struggle in a conciliatory way, even if he succeeded in obtaining a major victory. This led the Ambassador to renew the project of mediation; but the answer was still in the negative. The Prussian Government, thought Flahaut, was obviously influenced by the English refusal to Talleyrand, the news of which had just arrived in Berlin. Ancillon maintained that Nicholas could not agree to any intervention; he could not do so because of his own Russian subjects: the war had become too much a conflict between the Poles and the Russians. More, he believed, could be achieved if France were to advise submission to the Poles: in this case they would obtain their constitution and a pardon for the past. At the same time Ancillon referred to his former opinion that victory or submission was the only means to terminate the struggle.[88]

[87] Chad to Palmerston, 14, 20 July, F.O. Prussia, 12 July, Palm MSS.

[88] Flahaut to Sebastiani, 14, 21, 25 July. On the 25th Ancillon told him: "Dites leur (the Poles) qu'ils se confient à la générosité de l'Empereur; qu'ils se

The last encouraging comment of the French Ambassador was mad ein his report of the answer given by the "Preussische Staatszeitung" to a Warsaw newspaper which printed Skrzynecki's letter to Frederick William III, although the letter had not been accepted by the Berlin Chancery. Public opinion, he thought, compelled the Government of a great Power to explain its action to an authority it did not recognize.[89]

But his mission obviously was finished. The authorization to leave his post was dated 2nd August. His only tangible accomplishment was his intervention with the Prussian authorities on behalf of the messenger who was the bearer of the Polish dispatch from the Paris mission to the Government in Warsaw with the information of the French *démarche* in London. He also had very secret relations with some Poles in Berlin. A likely development of his action was made impossible by the news of the French failure in London. As it happened, he only succeeded in irritating the Prussian Government. Chad's influence was even less than that.[90]

In Vienna, too, the Polish insurrection was a subject of discussion and speculation. The British Ambassador referred to it again in the beginning of May. Two points were stressed: the inability of Russia to suppress the rising and her inability to incorporate the Kingdom of Poland within the Empire in view of public feeling in Europe. It was emphasized that although at the Congress of Vienna Prussia and Austria were opposed to the creation of a Kingdom of Poland, and although they did not readily admit the right of interference in this matter by other signatories of the Treaty, they were nevertheless opposed to the annexation of Poland by Russia now. Lord Cowley was sure that Metternich in particular felt very strongly the necessity of Poland continuing as a separate Kingdom.[91]

This information was accompanied by an account of the reports

---

soumettent à lui et *leur nationalité restera intacte;* ils obtiendront et leur constitution et le pardon du passé. Mais, je vous l'ai déjà dit: *cette question ne peut plus se décider que par la soumission ou la victoire.*" A.é. Prusse.

[89] Flahaut to Sebastiani, 31 July: "Il est remarquable de voir l'opinion publique forcer un des grands Cabinets à entrer dans des explications vis à vis d'un gouvernement qu'il ne reconnaît pas." A.é. Prusse.

[90] Flahaut to Sebastiani, 14, 18 July A.é. Prusse. On the rôle of his mission: J. Darcy, *L'ambassade de Talleyrand à Londres,* 1891, p. 20; Webster, I, 187. On Chad's lack of influence: William Russell to John Russell, 4 January, 1837, *Early Correspondence of Lord John Russell,* 1805–1840, 1913, II, 191.

[91] Cowley to Palmerston, 1 May, F.O. Austria.

from the Austrian officer at Diebitsch's headquarters, Col. Caboga, who represented the Russian army as completely demoralized. A later report attributed this opinion, *mutatis mutandis,* to Nicholas himself. The internment of Dwernicki's corps was, to a certain degree, a surprise. In their explanation the Austrian Government laid particular stress on the pressure put upon them by Russia; the relevant article in the "Oesterreichische Beobachter" was written to create the impression that only by the interposition of Austrian forces was a complete defeat of Dwernicki prevented, pursued as he was onto Austrian territory. It was known, however, that Dwernicki's men were poorly guarded and no particular care was taken to prevent their movements. The French Ambassador, at the General's request, intervened with Metternich on his behalf, while several officers of Dwernicki's corps visited the French Embassy. Marshal Maison assured them of the genuine interest of France in the Polish resistance. For the time being Dwernicki's case contributed to a *refroidissement* between Paris and Vienna.[92]

This question almost coincided in time with renewed French proposals for talks on disarmament. Metternich rejected the mere idea that France should take the initiative in such a matter; this was a task for the three Eastern Courts and since the July revolution he had been making efforts to bring about a renewal of their alliance, especially important at a time when affairs were so unsettled. It was the Polish revolution which accounted for his failure: the Eastern Powers were still trying in vain to suppress it. That is was Nicholas who refused an agreement on Polish affairs, Metternich, as it will be seen, preferred not to admit.[93]

On 26th June the special courier with the British Government's protest against the treatment of Dwernicki's corps arrived in Vienna. Before the British Ambassador called on Metternich, he informed Marshal Maison of the contents of his instructions. It so happened, therefore, that the Ambassadors visited the Austrian Chancery one after the other. Metternich was for a moment confused by the appearance of an Anglo-French agreement in this

[92] Cowley to Palmerston, 1, 21 May, F.O.; Maison to Sebastiani, 12 May, A.é. Autriche; Fagel to Verstolk, 20 May, R.A.; Prince Constantine Czartoryski to the Polish envoys in Paris, 21 May: The Austrian Cabinet considers war with France, sooner or later, inevitable. Cf. also Dutkiewicz, *Austria wobec powstania Listopadowego,* pp. 77 and 82-3.

[93] Metternich to Apponyi, 3 June, *Memoirs of Metternich,* V, 136; A. Zamoyski, *Moje Przeprawy* (My Adventures), Cracow 1906, p. 102. Cf. below, p. 280.

case. Yet he did not admit that Austria had committed any infringement of the law of nations; on the contrary, he told Maison that he had had difficulty in explaining to the Russians his disinclination to deliver men of Dwernicki's corps to them. A similar answer was given to Cowley. The Ambassador observed that no favourable policy towards the Poles might be expected from Metternich. The Chancellor did not conceal the arrival of a Polish agent who had already been the subject of Sir Robert Gordon's report from Constantinople, but did not admit that the agent had been seen by him; on the contrary, he maintained that he would not see the man. Lastly, Cowley wrote that he himself feared the extradiction of Dwernicki's men as Austria was anxious to win the gratitude of St. Petersburg; but he, too, knew that the interned soldiers were so poorly guarded that a considerable part of them had already escaped.[94]

When the news of the French intention to begin a diplomatic action in favour of the Poles spread in Vienna, Cowley remarked that, according to dispatches of the Austrian Ambassador in St. Petersburg, Nesselrode "within the period of a month repeatedly declared to Count Ficquelmont that the Emperor Nicholas rather than submit to foreign interference would abandon Poland altogether"; and in this determination, Cowley thought, Nicholas was supported and encouraged by Austria. Further in the same dispatch he noted that Paris reports "have given rise to various speculations among those who were favourable to the cause of Poland, such as "that resources in money and stores will immediately be afforded to the Poles by Great Britain and France and that a combined British and French squadron will be sent to the North Sea, thus placing the Emperor of Russia in a predicament with respect to Poland, similar to that in which the Sultan was placed with respect to Greece at the period of the battle of Navarino."

Cowley had seen a memorandum, written by a Pole, in which the sending of a fleet to the Baltic was represented as likely to achieve the deliverance of Poland without a continental war, whereas Austria would adopt a similar line of action to that which

[94] Maison to Sebastiani, 1 July, A.é. Autriche; Cowley to Palmerston, 2 July (3 dispatches), F.O. Austria. The agent, Count Andrew Zamoyski, was received in the utmost secrecy a few times during April and May. See Dutkiewicz, *Austria wobec powstania Listopadowego,* pp. 64 and 88. Dwernicki himself could have escaped easily.

she had followed in the Greek affair, and Prussia would follow Austria in that case. "They are not the avowed sentiments of Prince Metternich", continued the British Ambassador, "but some expressions have occasionally dropped from him which might lead one to believe that he himself thinks that Austria may be compelled to adopt this course and among his confidential friends he has allowed himself to say that were the Emperor of Russia to be placed in the situation I have mentioned he would not be subjected to harder conditions that those which he had contributed to impose on the Sultan." Cowley further attributed to Metternich the opinion that a defeat of the Russians in a subsequent battle "would secure the independence of Poland", while even after the taking of Warsaw Nicholas would not attempt "to re-establish his authority upon anything like the footing on which it stood previous to the breaking out of the insurrection." [95]

This was as yet far from his official utterances. The French Ambassador, who called on Metternich to sound him on the possibility of Austria joining the Anglo-French *démarche* in St. Petersburg in favour of the Poles, found the Chancellor disagreeably surprised at the still clearer appearance of a concert of the Western Powers in Polish affairs. The Emperor of Austria could never support insurrection, Metternich said; such a gesture would inevitably excite revolutionary movements in Hungary and Bohemia. Yet the Ambassador's impression was not entirely unfavourable, on this occasion he brought to the knowledge of the French Government

[95] Cowley to Palmerston, 20 July, F.O. Austria, also printed in H. Wellesley: *Diary and Correspondence of the First Lord Cowley* 1930, pp. 193 sq. The memorandum is preserved in the MSS of the Polish Library in Paris. The Paris mission received it from Prince Constantine Czartoryski, resident in Vienna, and forwarded it to Walewski; from the covering letter of the Prince it appears that another copy was sent with Cowley's dispatches to London. On the memorandum itself the Prince commented: "Cette notte est rédigée par un Polonais avec connoissance et même avec l'aide d'un personnage qui non seulement se trouve souvent consulté mais qui est au fait des secrets de la diplomatie en général et plus particulièrement encore de ceux du premier ministre d'ici." This description refers probably to Gentz. Further: "Le but de cette notte est de prouver à l'Angleterre que tout dépend d'elle, que n'ayant rien à craindre de la France, tout au contraire elle pouvait conter sur sa coopération. C'est à elle à prendre l'initiative. Qu'ici on ne demande pas mieux que d'avoir la main forcée à changer du système."
Unfortunately, the letter of the Prince being dated July the 16th, and Cowley's dispatch, the 20th, this assurance arrived in London at a time when the Government heartily regretted the cooperation of France, after the beginning of hostilities in Belgium and the unavoidable French intervention.

some "paroles remarquables" of the Austrian Chancellor, spoken to him with reference to Poland some weeks earlier. "Croit-on que, comme homme, je puisse être insensible à la vue de tant de courage, et pense-t-on que, comme ministre, je n'aimerais pas mieux avoir pour voisine une Pologne toujours bienveillante et toujours amie, qu'une Russie toujours envieuse et toujours envahissante?"

It is difficult to see what induced Metternich to make this statement after the Polish defeat at Ostrolenka, unless assumption is made that he also was impressed by the extent of the Polish movement and its echo abroad. It was assumed in Viennese circles that the Polish movement, whether ending in success or defeat, was bound to bring about important changes. Immediately after the interview referred to, the Chancellor sent couriers to Berlin and St. Petersburg which the French Ambassador believed to a part of the intercourse between the three Powers on concessions which they might be required to make in Polish affairs, though Metternich seemed to have inferred from Maison that the Western Powers would not push very hard; and as for obligations under the Treaty of Vienna, the Poles themselves had broken the engagement by their act in deposing Nicholas.[96]

All the evidence of interest shown by the Western Powers in Polish affairs presented a major problem to the Austrian Government, or at any rate it wanted at an early stage to impress this upon the Russian Government. Such was the aim of Metternich's dispatch to the Austrian Ambassador in St. Petersburg, dated July 8th. This document was written on account of Dwernicki's case, but refers to much more. Though it professes the principle that a sovereign and his revolted subjects cannot be treated on an equal footing, it nevertheless confirms at the same time by its contents that, in fact, circumstances compelled the Austrian Government to act as if they were. [97]

The final echo of the supposed Anglo-French *démarche* on behalf of the Poles, was the interview between Metternich and the

[96] Maison to Sebastiani, 21 July, 3 Aug., A.é. Autriche; Gentz to Prokesch, 17July, *Aus dem Nachlasse des Grafen A. Prokesch-Osten,* Vienna 1887, III, 41; Metternich's statement is printed in d'Haussonville, *Histoire de la politique extérieure du gouvernement français 1830–1848,* Paris 1850, I, 29 and repeated by Thureau-Dangin, *Histoire de la monarchie de juillet,* I, 167; but it is supposed to have been made early in 1831.

[97] Schiemann, *Geschichte Russlands,* III, 469.

British *Chargé d'affaires,* Mr. Forbes, in which they "exchanged" their respective refusals made to the French in London and Vienna.[98]

VI

The account of Flahaut's activity must be completed by a note on Palmerston's attempt to exercise pressure on St. Petersburg through Berlin and of Chad's findings in this affair. The subject of Poland was taken up by Palmerston in his private letter to Chad of 3rd May. He first repeated what he had said to Granville on the attitude of Prussia and Austria towards the Polish revolution. Palmerston realized that the Prussian Government could not adopt a sympathetic attitude with regard to the Polish movement; yet, especially as Russian arms did not seem likely to bring the contest to a speedy conclusion, Berlin might be desirous to help to bring about a reconciliation between Nicholas and the Poles. Chad was therefore instructed to avail himself of every opportunity to offer informal advice to this effect.[99]

Chad was not, as has been seen, the man to influence the Prussian Government on a point on which they felt very strongly. He had already reported on the inclination of Nicholas to treat with the Poles as soon as he had obtained a victory. After the receipt of the above letter Chad succeeded in mentioning the subject to Ancillon and obtained the impression that Russia had probably asked for assistance, the more so as Ancillon, unasked, told him that Russia would oppose a conference on disarmament fearing that such a conference would take up the subject of Poland. Some time later Chad learnt from the Hessian *Chargé d'affaires* that the King of Prussia accompanied his refusal of military assistance to his son-in-law "by an expression of his anxious desire to see the contest at an end, an expression calculated... to convey a hint of mediation which H.M.'s knowledge of the Emperor's feeling on this point prevents him from offering distincly." The French *Chargé d'affaires* was said not to believe in the story; but the Hessian gentleman added as certain, at any rate, "that the Prussian Government had caused it to be indirectly insinuated to the Polish Government

[98] Forbes to Palmerston, 2, 6 Aug., F.O. Austria.

[99] Palmerston to Chad, 3 May, Palm MSS. See above, p. 140 and Appendix IX.

that if they made an advance towards a reconciliation with Russia, they will be met half-way by the Emperor." [100]

This report corresponds rather with Nicholas' own letter to his brother and is probably based upon the incident in April, at the time of the beginning of the rising in Lithuania and of a somewhat more marked interest in Poland shown by the two Western Ambassadors. The embarrassment at St. Petersburg early in May is fully treated by the Netherlands Minister. His dispatch surveys some events of Polish history after 1793 (the second partition of Poland), and of the past campaign; stress is laid upon the excitement of popular feeling in Poland. "Si à l'époque de l'invasion de la Pologne," continued Baron Van Heeckeren, "la Russie s'était exprimée avec plus de modération et de convenance, si ses agens à l'étranger surtout n'avaient point exagéré sa puissance et affecté, par un patriotisme mal entendu, de déprécier la Pologne outre mesure, en représentant l'invasion de ce pays comme une promenade militaire; si le Maréchal Diebitsch lui-même, pendant son voyage à Berlin, n'y avait constamment tenu le langage d'une confiance illimitée, l'état des choses... paraîtrait moins fâcheux." Moreover, Van Heeckeren contrasted the noble attitude of the Poles with the obstinacy of the Emperor and pointed out that the interest of the Ambassadors of France and Britain did not stretch beyond the provisions of the Treaty of Vienna, insufficient as they were for the Poles; these were the factors responsible for the "guerre d'extermination" which had begun.[101]

The Western Ambassadors were not entirely idle. Mortemart, though not enthousiastic about the scale of Polish successes, noticed the obvious embarrassment of Nicholas. He again approached the British Ambassador and unfolded to him a plan of negotiations to be effected through the good offices of the Prussian Court, whereby the King of Prussia would ask the Poles to submit and guarantee to them equal representation in the negotiations with Nicholas. Lord Heytesbury, whose instructions were "à ne laisser échapper aucune des occasions" for a confidential offer of mediation, as noted by the Netherlands Minister, "paraît assez goûter ce projet, mais ni l'un ni l'autre des deux Ambassadeurs ne sait encore par quels moyens le faire agréer à l'Empereur." The Polish

[100] Chad to Palmerston, 9 May, 8 June, F.O. Prussia, 18 May, Palm MSS; see above, p. 130-1 and 176-210.

[101] Van Heeckeren to Verstolk, 13 May, R.A.

army would always remain a difficulty, as it appeared unlikely that Nicholas would agree to keep it in being, while the Poles could hardly be induced to renounce their only force. The French Ambassador had yet other arguments: that the war party in France would cause Périer's cabinet to fall in consequence of the prolongation of the war in Poland; hence the necessity for Russia not to refuse to negotiate. If this attitude impressed St. Petersburg circles, Mortemart himself was not in the least assured of his success. His reports from the same period are couched in an apologetic tone with regard to the Poles: he was glad that the rising was holding out in Lithuania and Volhynia, and that Dvernicki's defeat "n'a nullement ralenti le zèle des insurgés." On the other hand, he feared the consequences for the Poles of the expected appointment of Paskevitch.[102]

The French Ambassador was also uneasy about his collaboration with his British colleague. It has been seen on previous occasions how much the French Government insisted that initiative in Polish affairs should come at least in appearance from London. Neither was Heytesbury, however, the man for the situation nor were the relations between Paris and London such as to lead one to expect frank co-operation, nor, in the last instance, were the French Ambassador's own views likely to bring about a positive result. Mortemart's attempt, described above by the Netherlands Minister, took a different form under the Ambassador's own pen. His dispatch of 4th June begins by a few flattering remarks on Heytesbury. He then notes that Heytesbury approached him one day (obviously for the second time, since more than a month had elapsed between the date of the dispatch and Heytesbury's first interview with Nesselrode) on the subject of Poland, wanting to know his instructions. The French Ambassador had not yet received his instructions of 17th May, which allowed him to act more freely, according to circumstances. He said that he would be happy to join in Heytesbury's own efforts; but Heytesbury had as yet nothing in his mind. "Je lui proposai quelques insinuations à mettre en avant avec propos", wrote the French Ambassador, "il m'approuva". But it appeared that the British Ambassador had only once mentioned the subject to Nesselrode, and never said a word about it to the Emperor, in accordance with the little interest

[102] Mortemart to Sebastiani, 4, 25 May, A.é. Russia; Van Heeckeren to Verstolk, 18 May, R.A.

that, in Mortemart's opinion, England took in those affairs. "J'ai donc été fort étonné", continued Mortemart, "de lire ensuite que V.E. m'engage à entrer dans les vues de Ld Heytesbury et de donner une adhésion pleine de réserve aux efforts qu'il peut tenter. Je le répète, nos vues sont les mêmes, mais jusqu'à ce moment, seul j'ai fait des efforts dont l'Angleterre veut bien recueillir les fruits, sans, en aucun cas, rien compromettre. J'ai cru devoir indiquer sur le champ à V.E. ma position à cet égard et la mettre à même de prévoir la valeur de la co-opération de l'Angleterre dans l'affaire de Pologne." The indignation of the Duc de Mortemart is still better explained when it is added that he claimed to talk of Poland almost daily to Nesselrode as well as to Nicholas.[103]

In many other instances, however, the fear of an Anglo-French agreement was supported by the appearance of a concerted action. The subject of mediation in favour of the Poles continued to be the diplomatic news from the Russian capital during the month of June. Mortemart made his official *démarche* on the ground of his instructions of 15th May, in which the cholera epidemic served as a pretext for the French Government to try to bring an end to the hostilities. The Ambassador himself, and his British and Dutch colleagues observed that the offer was ill received; Mortemart thought that Nicholas absented himself from military parades to avoid meeting him. Heytesbury was perhaps more cautious in his comment: circumstances might perhaps compel the Imperial Government to treat; yet they could not think of agreeing to the independence of Poland because this would have been followed by the loss of all the former Polish provinces. Van Heeckeren reported that the Russian answer employed measured expressions; still he qualified it as "refus coloré mais formel." On the other hand, he was convinced that a Prussian offer of mediation would be looked upon with favour, and that the Prussian Government was renewing its efforts at Warsaw for that purpose. The situation in Lithuania was deteriorating daily. The insurgents were planning an attack upon Libau, "où doivent arriver de la France et de l'Angleterre des vaisseaux qui leur apportent des munitions de guerre." Finally, Podolia and Volhynia seemed by no means pacified.[104]

[103] Mortemart to Sebastiani, 4 June, A.é. Russia. There is no instruction for Mortemart dated May 17, and that of the 15th does not correspond with the contents as described by the Ambassador.

[104] Cf. the Prince of Orange to Mme Lieven, 31 May: "Je vois par les dépêches de M. de Heeckeren que le Duc de Mortemart et Lord Heytesbury se donnent

It must be noted that these reports were made at St. Petersburg within a fortnight of the arrival there of the news of the victory at Ostrolenka. The commander of the victorious army, Marshal Diebitsch, fell a victim to cholera, or, according to various confidential reports, to poison. His successor, Paskevitch, whose reluctance to take over the command had been known before, went "souffrant, désolé et sans confiance", taking the sea route which would lead him through Prussian territory (as a passage through Lithuania appeared unsafe); Mortemart noted that a possible exemption from quarantine regulations for the Russian commander would constitute a strong proof of the partiality of the Prussian authorities. A new levy of troops was announced in St. Petersburg and the British Ambassador reported difficulties in recruiting.[105]

In the beginning of July the French Ambassador observed uneasiness even in the relations between Berlin and St. Petersburg. There were confused reports about Paskevitch, but in the end London knew that he was kept in quarantine. Pozzo's reports which now penetrated into the Russian capital, increased the fear of revolution in France, but instead of making the Imperial Cabinet more tractable – as the corresponding reports did in the case of the London conference – they resulted in an increase of anti-French feeling which only the Polish troubles prevented from fanning into an outbreak. On the other hand, Austrian military preparations at the French frontiers were denied. From Polish quarters it was learnt that Nicholas had written to the Emperor of Austria to tell him that as soon as he had obtained some military advantages he would treat with the Poles. Mortemart was sure that only a survey of the diplomatic correspondence of all the Powers concerned could explain various contradictions in the current news.[106]

beaucoup de mal pour parvenir à faire terminer les affaires de la Pologne par une négociation et qu'ils veulent que le Roi de Prusse soit l'intermédiaire entre les Polonais et l'Empereur, mais ils ne savent pas comment s'y prendre pour faire accepter ce plan à l'Empereur qui est déterminé plus que jamais de finir par les armes." Lie MSS. Mortemart to Sebastiani, 22, 25, 29 June, A.é. Russie; cf. above, p. 170. Heytesbury to Palmerston, 15 June, F.O. Russia; Van Heeckeren to Verstolk, 15 June, R.A.

[105] Heytesbury to Palmerston, 31 May, 27 June, F.O. Russia; Chad to Palmerston, 22 June, F.O. Prussia; Mortemart to Sebastiani, 18, 29 June, A.é. Russie.

[106] Mortemart to Sebastiani, 2 July, A.é. Russie; Greville noted on 8th July that Paskevitch was allowed to pass through Prussia, but on the 13th observed that the Marshal was, after all, detained. *Greville Diary,* II, 160 and 163; Barzykowski, IV, 126.

Soon after this report cholera broke out in the Russian capital, accompanied by popular riots. During the first two weeks of July the epidemic was almost the only topic in diplomatic communications from St. Petersburg. The diplomatic corps were abandoned entirely to themselves. Both Heytesbury and Mortemart did not fail to observe that those two weeks furnished a good deal of convincing proof of the weakness of the Russian Empire; Heytesbury for a moment considered leaving St. Petersburg, as all communications were broken.[107]

This situation immediately preceded the arrival of new instructions for the French Ambassador, dispatched in consequence of the French proposal to the British Government made on 7th July, and since then widely known in other capitals. The diplomatic corps expected negotiations with regard to Poland. Heytesbury awaited new instructions and believed that they would formally order him to intervene jointly with the French Ambassador, in such a way as to exclude any delaying measures on the part of the Russian Government. The Netherlands Minister thought that Russia would be unable to refuse the offer and thereby risk a rupture with the other Powers. Recent events proved that the fear which Russia used to inspire in the other Powers had entirely disappeared. "Il est donc naturel de croire", wrote Van Heeckeren, "qu'un grand changement se fera désormais remarquer dans le ton que prendront les autres cours vis-à-vis de la Russie, et dans celui qu'elle devra mettre elle-même dans ses réponses." Also about this time arrived Matuszewic's report of his interview with Palmerston in which the latter had told him that Poland and Lithuania were lost to Russia; the Austrian Ambassador in St. Petersburg lost no time in informing his Government of this opinion.[108]

In fact, Mortemart alone received new instructions. He had already on several occasions spoken to Nesselrode in the same spirit, and now felt strengthened to pursue his own line. The Vice-Chancellor always had the same reply: the Emperor would see his way to grant mercy to the Poles. For his part Nesselrode ap-

[107] Heytesbury to Palmerston, July, F.O. Russia; Mortemart to Sebastiani, 6, 12 16 July, A.é. Russie; Van Heeckeren to Verstolk, 19 July, R.A.

[108] Van Heeckeren to Verstolk, 19 July, R.A.; A. Zamoyski, *My adventures*, I, 58: (written about August 10) "A courier from St. Petersburg brought Ficquelmont's report that Matuszewic wrote of Palmerston telling him 'que désormais ni la Pologne ni la Lithuanie ne pourront appartenir la Russie'." Cf. above, p. 169.

peared to rely on Heytesbury's opinion that England would not press the Polish case; therefore the Russian Minister was not prepared to give way to French insistence alone. As for Mortemart, he did not want to give up; leaving Nicholas undisturbed in his Polish policy would amount to the annihilation of Poland, "car la haine est à son comble dans le coeur du souverain." Reports concerning Western intervention in Polish affairs in Vienna and Berlin poured into St. Petersburg, but they still failed to alter the attitude of the Government.[109]

Some slight progress was achieved by the French Ambassador on the 28th. In view of his new instructions he signalled to Paris his intention to speak plainly on Poland, and he tried even some sort of menace. He asked Nesselrode for a definite declaration with regard to Poland, written on the Emperor's instructions, as he himself wanted to take it to Paris on his return, his special mission being about to end. The Ambassador was sure that this proposition greatly impressed the Vice-Chancellor as an evidence of a possible Russo-French rupture; the presence of Mortemart, an Ambassador from the Bourbon Court, was greatly appreciated in the Russian capital. But this interview had to remain sterile in its consequences: soon after this a Russian courier from London brought the information that Britain refused to support the French offer of mediation. The Vice-Chancellor hastened to send word to Heytesbury and thus it became known to the French Ambassador.[110]

Mortemart realized that his mission, as far as Poland was concerned, had come to an end, and moderated his tone. He did not fail to note, however, that the suppression of Louis Philippe's speech at the opening of the French Chambers gave rise "aux interprétations les plus favorables à la cause polonaise"; these were not shared by himself, but the English refusal he described in terms showing anxiety of St. Petersburg with regard to Western interest in Poland.[111]

[109] Mortemart to Sebastiani, 23, 27 July, A.é. Russie; Heytesbury to Palmerston, 22, 27 July, F.O. Russia.

[110] Mortemart to Sebastiani, 30 July, A.é. Russie.

[111] Mortemart to Sebastiani, 10 Aug., A.é. Russie; Van Heeckeren to Verstolk: "Le ministère anglais avait déclaré que quand même il aurait été d'avis d'offrir sa médiation aux *deux* partis belligérants, il n'aurait jamais choisi pour le faire le moment où les malheurs de tout genre tombaient sur l'Empire." R.A. The word "deux" underlined by the writer. The first part of this text corresponds clearly with Mme Lieven's letter; the reference to misfortunes, however, seems to reflect rather the feeling of the writer.

In August disturbance and later open mutiny in military colonies additionally embarrassed the Imperial Government; at this time news about hostilities in the Low Countries and of the French entry into Belgium began to arrive in the Russian capital. The Netherlands Minister wrote that, from then on, the French Ambassador adopted an aggressive attitude. Lord Heytesbury noted that the warlike tone of the French newspapers made the ending of the war in Poland imperative, and added news about a new levy of troops throughout the Empire. Mortemart found out that Heytesbury took his explanations of the measures undertaken by the French Government in Belgium at their face value and this greatly contributed to calm down the feeling at St. Petersburg. The Austrian Ambassador and the Swedish Minister were "les brouillards du Conseil".[112]

The French Ambassador had his own views about the effect of his diplomacy with regard to the intervention in Belgium. With this news coincided his leave-taking audience of the Emperor and Nicholas was interested to know whether, after the withdrawal of the Dutch, the French Government would be able to keep their promise to withdraw their army as well. He readily accepted the Ambassador's affirmation, and, though in a somewhat different form, caused it to be known to Pozzo at Paris. "L'Empereur a été si coulant, si amical et si aimable que je commence à croire au bruit qui se répand que les affaires vont mal en Pologne." [113]

With this impression the Duc de Mortemart left St. Petersburg for France on 27th August.

[112] Heytesbury to Palmerston, 3, 13, 18 Aug., F.O. Russia; Van Heeckeren to Verstolk, 20 Aug., R.A.; Mortemart to Sebastiani, 17, 20 Aug., A.é. Russie.

[113] Mortemart to Sebastiani, 20 Aug., A.é. Russie; Nesselrode to Pozzo, 24 Aug.: "L'Empereur a dit à l'Ambassadeur de France "que le Roi Louis Philippe n'étant plus le maître chez lui, le language de ce prince ne saurait ni le blesser ni l'atteindre." Martens, *Recueil,* XV, 138. The remark is said to concern chiefly "la mention polonaise" in the speech to the Chambers.

CHAPTER V

THE "TEN DAYS CAMPAIGN", THE FALL OF WARSAW AND THE TREATY OF 15 NOVEMBER 1831

The month of August, 1831, seemed to offer better prospects for peace at least in the Low Countries. The diplomatically troublesome Belgians were engaged in enjoying the first fruits of their apparently assured independence, giving a warm reception to the new King, who crossed the French border into Belgium on 17th July.

Though the military situation in Poland did not warrant an unequivocal hope for victory on the part of the Poles, they, knowing nothing of what had passed on the subject of intervention in London, were still expecting some diplomatic steps in their favour. Little was known abroad about the real strength of the contesting parties, and the Eastern Powers were by no means easy about France's intentions with regard to Poland. Neither Vienna nor St. Petersburg were satisfied with the military situation in Poland; from Berlin Chad reported that the Russian passage over the Vistula, officially acclaimed by the "Preussische Staatszeitung" as a brilliant operation, was confidentially described by Prussian officers as very unskilful.[1]

Louis Philippe's speech at the opening of the French Chambers on 23rd July was subjected to various comments, which showed that no quarter was satisfied with the expressions used with regard to Poland; those who knew or believed that the French Government had no intention of intervening in Polish affairs could with reason fear that the days of Périer's cabinet were numbered, especially as it had no successes elsewhere to report.

There was no unanimous support in London for Leopold's candidature to the Belgian throne, but hostile disapproval would be the better word to describe the feeling in Paris in this respect. However, Leopold was accepted in Brussels. And though the XVIII Articles were rejected by the King of the Netherlands, who more-

[1] Chad to Palmerston, 4 Aug., F.O. Prussia.

ver declared he would consider Leopold an enemy should he enter Belgium, little attention was paid to views originating in The Hague.

I

As early as the beginning of July the opinion, spread from the Berlin Court, which was always closest to that of The Hague, was that the King of the Netherlands would neither accept the XVIII Articles not consider a financial indemnity for Luxemburg.[2]

King William I had hoped that the Belgians would refuse the new propositions, thus placing him in a more favourable position. But they accepted. In his speech to the Belgian Congress on 5th July the Belgian Minister for Foreign Affairs, Lebeau, pleading for acceptance, included some totally unwarranted interpretations, as if the whole of Limburg had been given to Belgium. As this was not palatable to the King of the Netherlands, "il faudra que le Roi Guillaume fasse la guerre", he said. This speech was greatly disapproved of in London. Though William I, at the request of Bagot, delayed official announcement of his decision until he heard of the reactions of the Conference to Lebeau's speech, it was obvious that this speech sealed his determination. It must be added, however, that the news that the departure of Prince Leopold for Brussels had been agreed upon before the Conference learnt the views of the King of the Netherlands could not but provide further grounds for this decision.[3]

The Conference, or rather Palmerston, was determined to send Leopold to Brussels as soon as it heard of the Belgian Congress's favourable vote. The Prince himself felt by no means comfortable about it and would have preferred to wait for King William's reply; but Matuszewic, on behalf of the Conference and more particularly speaking for the Eastern Powers, assured him that he would be recognized as King, even if the King of the Netherlands refused to subscribe to the Articles. This refusal arrived in London on the 20th July, when Leopold had already set foot on Belgian soil.[4]

[2] Chad to Palmerston, 5 July, F.O. Prussia; Flahaut to Sebastiani, 9 July, A.é. Prusse. Cf. above, p. 162.
[3] Bagot to Palmerston, 11, 12 July, Bagot MSS, 15 July, F.O. Holland.
[4] Falck to Verstolk, 27 June, R.A.

The Prince left London without entertaining great hopes for his future. To Wellington he is reported to have said that his Belgian kingship would last only as long as the Cabinet of Casimir Périer in France. He took leave of the Lievens as well, who assured him that if an Orange restoration was impossible, Leopold's person would offer guarantees of stability, which would be appreciated by the Emperor of Russia too. Mme Lieven seized the opportunity to mention the presence of Załuski in Brussels and to impress upon the Prince that he would also be judged in St. Petersburg by his attitude towards Polish agents.[5]

Palmerston, however, believed that he had accomplished a successful step. He saw the uncertainty drawing to a close. Leopold's taking over of the government in Brussels was considered by him as a final blow to the policy of the King of the Netherlands, who in this way was deprived of the means of keeping affairs unsettled as long as the Belgian kingship remained dependent on his consent, and he cheerfully described the King's anger as he imagined it. The Minister was confident that William I would not dare to recommence hostilities. And Palmerston did not care whether the Eastern Powers would keep to their promise to recognize Leopold. Recognition only by England and France was sufficient, and perhaps even advantageous for Leopold with regard to his own subjects, for it made his appearance in Brussels less that of a "child of the Conference". "It may not be amiss", he wrote to the British Ambassador at The Hague, "to hint this to the Dutch who probably think they have a great hold over him by the influence they might exercise over Russia if not also over the other two, to delay their acknowledgement of him".[6]

Bagot's reports from The Hague, however, continued to describe warlike feeling in Government circles. "They feel perfectly at liberty", he wrote on 25th July, "to go to war if they please & I *know* that several of the Ministers now reproach themselves with not having counselled war immediately upon the receipt of the last propositions.... (i.e., the XVIII Articles)." The Ambassador

[5] *Greville Diary,* II, 180, 5 Aug. 1831: "Matuscewitz told me that he [Leopold] went on his knees to Palmerston to send somebody with him who would prevent his getting into scrapes, and that Falck and Talleyrand, by far the best heads among them, had both predicted that Leopold would speedily commit some folly, the consequence of which might be irreparable."; Mme Lieven to Nesselrode, 15 July, Lie MSS.

[6] Palmerston to Bagot, 25 July, Bagot MSS.

heard of the King's enquiry of the Prussian observer at Dutch Headquarters as to whether Prussia would come to the King's assistance in the event of hostilities against the Belgians. On the 31st Van Zuylen left The Hague for London, bearing the King's answer to the proposal for new negotiations, and on the same day the Prince of Orange left the capital for Breda, reported – but Bagot was not sure – to have been appointed Commander-in-Chief of the Army.[7]

On 2nd August the Dutch Army crossed the Belgian frontier. It soon appeared that the Belgian army was retreating everywhere.

King Leopold, who learnt of the Dutch attack during festivities at Liège in the afternoon of the 2nd, instructed Lebeau, the only person of distinction at his side, to write to the Belgian representative in Paris ordering him to request the immediate armed assistance of France, in virtue of her being a party to the agreement guaranteeing Belgian neutrality. The French Government hardly waited to receive this request. Their representative in Brussels, Belliard, who, incidentally, had just returned from the French capital, went on his own responsability to Antwerp and succeeded in getting the Dutch commander of the Antwerp citadel to withdraw his notice of the cessation of the suspension of hostilities. In Paris, at the first news of the invasion – on the 4th – Marshal Gérard was appointed Commander-in-Chief of the "armée du Nord", and he left Paris that night with detailed instructions for the occupation of Belgium.[8]

The Belgian Cabinet, however, unaware of the extreme danger, did not countersign the King's *démarche* from Liège, and instructed Le Hon in Paris to request that the French army be held in readiness at the Belgian frontier in anticipation of the Belgian King's orders. When Le Hon communicated this request to Sebastiani the French Minister replied that the French Army would enter Belgium even if it had to open the gates of Mons "à coup de canon". Accordingly, by special orders from Paris, issued on the 5th, the entry of the French troops was fixed for the morning of the 7th, even before Marshal Gérard received the letter from the Belgian Minister for Foreign Affairs, in which the latter requested notifica-

[7] Bagot to Palmerston, 25, 29, 31 July, Bagot MSS.

[8] A. Martinet, *Léopold Ier et l'intervention française en 1831*, 1905, pp. 104-5, 124 sq., 131. Lebeau ceased to be Minister on Leopold's arrival, and only after the news of the Dutch invasion was he appointed Minister without Portfolio.

tion of Gérard's headquarters to which Leopold's orders, "qui doivent nécessairement précéder votre entrée en Belgique," [9] could be sent.

In the meantime, the news was reflected in diplomatic circles. Bagot, after sending his first account of the beginning of hostilities on the 2nd, wrote the next day that public opinion in Holland gave its whole-hearted support to this act of the King, though it was not unaware of the grave implications. It was by Bagot's letter that Palmerston was first acquainted with this extraordinary news.

The French Cabinet resigned the day before Paris learnt of the Dutch invasion, but now immediately resumed office. No time was lost in making French intervention acceptable to the Conference Powers. After the appointment of Gérard the French Minister for Foreign Affairs summoned a conference of the representatives of the Conference Powers and explained to them the character of the French intervention which had for its only object the protection of Belgium from the Dutch assault. When they objected to this measure, which, they said, should have been decided upon by the London Conference, the Minister answered, that the time available had not allowed such a procedure. At the same time he instructed Count La Rochefoucauld, the French *Chargé d'affaires* in The Hague, to request the immediate withdrawal of the Dutch troops from Belgium, and, should this request be refused, to ask for his passport and to leave his post. A copy of this dispatch was sent to Talleyrand to be shown to the British Ministers.[10]

The Dutch notification of the re-opening of hostilities was lodged at the Foreign Office probably only a little later than the actual crossing of the Belgian frontier by the troops of King William I, but its contents might have been known before. Its bearer, Baron Van Zuylen, arrived in London on 2nd August late in the evening and immediately informed Palmerston of his intention to call on him early in the morning. Accordingly, early in the morning of the 3rd he went to Palmerston's private residence in Stanhope Street.

[9] H.T. Colenbrander, *De afscheiding van België,* 1936, p. 105; Martinet, p. 138 and 151; Sol to Belliard, 7 Aug., "6 heures du matin" (from Brussels) "Le commandant de place de Mons n'a pas reçu d'ordres pour la réception des troupes françaises qui paraissent être en marche à l'heure qu'il est." Belliard, *Memoirs,* II, 403.

[10] Bagot to Palmerston, 3, 4 Aug., Bagot MSS; Granville to Palmerston, 3, 4 Aug., F.O. France; Sebastiani to La Rochefoucauld, 4 Aug., copy in A.é Angleterre, printed Ged., II, 237.

On handing over his dispatches to the Foreign Secretary, in the presence of the Austrian plenipotentiary Wessenberg, he uttered a single sentence: "My lord, vous verrez que nous sommes en guerre avec Léopold." Palmerston, however, being in a hurry, neither opened the dispatches nor asked for any explanations. "On Thursday evening [the 4th] when the advance of Dutch troops was known, Lord Grey and Lord Durham went to Stanhope Street for Lord Palmerston, and when they went to the Foreign Office, Lord Palmerston opened Zuylen's letter which communicated the intention of the Dutch Government." [11]

This singular incident delayed any measures the British Government might have wished to adopt in order to prevent both the Dutch advance and the French intervention, and the British Ministers felt outraged at being taken by surprise, while the situation in itself was serious enough. That is why "if the Dutch bombard Antwerp I will not leave a Dutch ship on the sea" was the first reaction of Grey; "Is the King of the Netherlands mad or who has led him to such insane proceedings?" Palmerston burst out.

In public Palmerston tried to be optimistic about peace, but he regretfully reminded Bagot of his having foretold the French entry – the first dangerous and inevitable consequence of hostilities in the Low Countries, while more were perhaps still to follow. To Van Zuylen he angrily wrote about the sending of the British Fleet to the Downs.[12]

But the Dutch felt they had some support from the Eastern members of the Conference, who fully aware of the serious situation took it upon themselves to represent it as the consequence of arranging Belgian affairs without due regard to the interests of both parties concerned.[13]

Nor were the French long in recognizing the favourable circumstances thus created. Talleyrand reported a difficult session of the Conference, when the drafting of a communication to William I, asking him to withdraw his troops from Belgium, met with very

[11] H. van Zuylen van Nyevelt, *Memorandum* (privately printed), 1853, pp. 10 sq.; Ellenborough Diary, 6 Aug., Ellenborough MSS.

[12] Grey to Mme Lieven, 12 Aug., *The Lieven-Grey Correspondence*, II, 267. The date is erroneous as the words are quoted in Mme Lieven's letter to A. Benckendorff, 5 Aug., Lie MSS; Palmerston to Bagot, 5, 6 Aug., Bagot MSS.; "Lord Palmerston rather surprised some of us by saying that he thought the peace of Europe by no means hopeless." Hobhouse, *Some Account of a Long Life*, II, 131.

[13] Falck and Van Zuylen to Verstolk, 4 Aug., R.A.

reluctant support from the Russian delegation; and on the same day Sebastiani pointed out in his dispatch to London that there should be advantages resulting from the expedition to the Low Countries, namely, the demolition of the fortresses on the French frontier and the recovery of territories lost by France in 1815. This did not prevent him from sending round an official note to the representatives of the Four Powers in Paris in which the re-establishing of the armistice was represented as the only object of the French move.[14]

Not only Britain but also the other three Conference Powers were offended, because Leopold's request had been directed, in the first instance, to France only; were the French action to be presented as a measure adopted by the Conference, Prussia would at least have to be included also. Sebastiani, however, almost ironically replied that Leopold's delicacy would prevent him from asking the King of Prussia's assistance against the latter's brother-in-law, whilst "the contemporaneous operation of a French and Prussian army might give rise to a dispute about the occupation of places prejudicial to the maintenance of general peace." In a different way this view was put forward in London. Representatives of the Eastern Powers, notably Bülow, feared that the French by acquiring, even for a short time, the control over the land between the Rhine and the Meuse, would command "un moyen d'action sur la Prusse, relativement à la question Polonaise." In this connection they expressed their regret that King William I did not propose the occupation of Antwerp to Britain and that of Maastricht to Prussia "à titre provisoire et pour faire compensation aux forteresses limitrophes de la France." Van Zuylen observed that "Prussia found herself in a serious situation, but as Bülow was constantly professing the weakness of his Court, any half-measures would satisfy him." There was, therefore, every prospect of success for France.

The Dutch plenipotentiaries were soon given supplementary evidence of the regret felt by the friends of Holland on account of the hostilities. Falck and Van Zuylen were satisfied that their friends in London – both English and foreign – would not concur with the

[14] Sebastiani to Talleyrand, 5 Aug., Talleyrand to Sebastiani, 5 Aug., A.é. Angleterre; Sebastiani to Granville, 5 Aug., 9 p.m., enclosing a Note of the same date, and a copy of Le Hon's letter to Sebastiani, dated "4 Août, à 8 hrs du matin". F.O. France.

view that the other Powers had not been adequately prepared to expect hostilities on the Dutch side in virtue of previous notes from The Hague. But opinions they heard indicated that the King had hardly chosen the right moment. Firstly, even if Warsaw had already fallen – an event of international importance - time was required for the change thus brought about to have any effects in international politics. Secondly, assuming that there would be war, it was in Holland's interest to keep its forces intact until then.[15]

II

On 7th August reports of the supposed entry of the French troops into Belgium reached London. Talleyrand witnessed a wave of anger and apprehension. "Deux conseils du Cabinet ont été tenus dans la journée", he wrote. In his next dispatch, however, he discussed at some length the prospects opened to France by the situation which had now arisen. The current Protocol, No. 31, drawn up hastily though with extreme difficulty in order to legalize the French expedition as a measure approved by the Conference, provided that neither Maastricht nor Venlo should be occupied by Marshal Gérard's army. It was clear that Prussia intended to occupy the two fortresses and Talleyrand saw in this a means to weaken Prusso-Dutch relations. Yet the most important result of frequent communications between Paris and Berlin in consequence of the French forces operating in the vicinity of the Prussian frontier would be the creation of a situation "qui permettrait de demander à la Prusse d'être neutre dans le grand débat élevé entre la Pologne et la Russie." The Poles should then be able to draw supplies from abroad. "Il paraît que ce n'est que cela qui leur manque pour repousser les Russes dans leur territoire." And lastly, the opposition in France would be deprived of any pretext for accusing the Government of "tiédeur" in the Polish question.[16]

The entry of French troops into Belgium, already warmly acclaimed in Brussels, had not as yet materialized. To his utmost astonishment Marshal Gérard was informed by the Belgian Govern-

[15] Granville to Palmerston, 5 Aug., F.O. France; Falck and Van Zuylen to Verstolk, 6 Aug., 9 Aug., R.A.

[16] Talleyrand to Sebastiani, 7, 8 Aug., A.é. Angleterre. The Protocol is printed in B.F.S.P., XVIII, 824.

or of Hainaut, on the morning of the 7th, that the latter had not been authorized to allow French troops into Belgium. The Marshal immediately referred to Paris for further instructions. Later during the day the Belgian Governor's instructions were confirmed by Belliard himself. The 7th and the 8th of August were thus spent in correspondence. The French Cabinet did not alter their determination: Gérard was ordered to enter Belgium, and to by-pass all places where authorities were unwilling to admit his troops, and to get in touch with the Dutch forces as soon as possible. He did not need, however, to await the arrival of these instructions. The news of the total defeat of General Daine, commanding one of the two main Belgian corps, made the Belgian Cabinet request the intervention of French forces. The first of Gérard's regiments crossed the frontier at 5.30 a.m. on the 9th of August.[17]

Whilst diplomacy was engaged in preventing France from profiting by favourable circumstances, the Dutch Army under the Prince of Orange advanced into Belgium. On the 8th General Daine was defeated at Hasselt. Belliard, not knowing of the Belgian Cabinet's decision, sent a courier to Gérard, with the request for an immediate entry into Belgium. On the 9th the Prince of Orange's headquarters were at St. Trond and his troops marched on Aerschot. In the night of the 11th the advance of Duke Bernhard of Saxe Weimar's army on Brussels was halted for a time, their commander agreeing to Belliard's request to him to seek new instructions from the Prince of Orange before continuing his advance. By the afternoon of the 12th the campaign was finished, the battle of Louvain being interrupted by the intervention of Sir Robert Adair, the British Ambassador, who, having only just arrived, exposed his life to danger by appearing in the front line. King Leopold, as was soon known, narrowly escaped being taken prisoner.[18]

All the pressure which the Conference Powers brought to bear upon the Dutch Government was not sufficient to make it order

[17] Sol to Belliard, 6 Aug., "à 8 heures du matin": "la nouvelle de l'arrivée de l'armée Française en Belgique est répandue dans toute Bruxelles et a été accueillie avec le plus grand plaisir." Belliard, *Memoirs*, II, 401; Martinet, pp. 167, 174-5, 196, 199.

[18] Colenbrander, *De afscheiding van België*, p. 105; Martinet, p. 195; J. Collin de Plancy, *L'histoire de Léopold, premier roi des Belges*, 1835, p. 162; Bagot to Palmerston, 15 Aug.: "The Duke of Saxe-Weimar is said to have been within an inch of taking Leopold." Bagot MSS.

the retirement of its troops before their direct contact with the army of Marshal Gérard, which, as has been seen, did not enter Belgium until the 9th. It was not until some days had passed that the diplomatic corps at The Hague realized this somewhat curious determination of William I, and the French *Chargé d'affaires* sent a courier to Marshal Gérard requesting him to get in touch with the Prince of Orange as soon as possible. Two days, according to Bagot, would have elapsed before the courier sent on the 10th, could have reached the French headquarters. In the meantime, however, King Leopold agreed to Sir Robert Adair dispatching Lord William Russell, attaché at the British Embassy, to Dutch Headquarters with Adair's letter informing them of the arrival of the French army in Wavre. But it was not until Adair himself appeared before the Prince of Orange that he agreed to the suspension of hostilities.[19]

Neither Paris nor London was as yet aware of the extent of the Belgian discomfiture. Orders sent from Paris on the 11th instructed Gérard not to harass his soldiers by precipitated marches, and, indeed, the French commander himself, knowing little of the real course of operations, was only too glad to allow his newly recruited soldiers to rest already in Mons. Palmerston was sure that the menaces held out to The Hague Cabinet would make them order the retreat as soon as the first news from London reached Holland. The Minister cheerfully wrote about the fading prospects of the Dutch to entangle the Four Powers in a general war against France. "This was not a bad speculation", he observed, "but like most of the Dutch plans, it comes a little too late; if they had moved before Leopold became King & whole Belgium was without a Government and chief, their plan might have answered." As has been seen, little had been needed at the time to make this a reality.[20]

Not even at The Hague did the diplomatic corps realize what had really happened. The representatives of the three Eastern Powers sent embarrassed and disapproving reports to their Courts, and the Prussian Minister was very active in trying to dispel any hopes William I might have of Prussian assistance; so was the King of Prussia with regard to the impression that might have arisen in

[19] Bagot to Palmerston, 9, 11 Aug., F.O. Holland, 9, 13 Aug., Bagot MSS; Martinet, pp. 257-263.

[20] Martinet, pp. 216, 258; Palmerston to Bagot, 9 Aug., Bagot MSS; cf above, p. 201.

the mind of his Imperial son-in-law about the improvement at The Hague. On the contrary, "much as he was right in complaining," wrote Frederick William III to the Empress Nicholas, at the close of the campaign, "his decision to open hostilities was incomprehensible." [21]

In the meantime, almost three days before the actual cessation of hostilities, the Netherlands Government announced its intention to withdraw the army from Belgium. It must necessarily seem ironical for Minister Verstolk to declare that the campaign in Belgium might have been a purely domestic affair, but "concevant les complications qui pourraient résulter de l'entrée des troupes françaises en Belgique" the Dutch troops would be ordered to retire. This declaration relieved the tension in London; Palmerston himself acknowledged his fault and told Van Zuylen that he had understood "la guerre" in his statement of 3rd August as "guerre morale", and therefore made no comment. The Netherlands Ambassador was abundantly congratulated on the victories of the Prince of Orange's army. The British Cabinet could assure Parliament that on the withdrawal of the Dutch forces the French would follow the same course. It did not appear, however, that this was their intention.[22]

On 12th August Palmerston learned of the conversation between Bülow and Talleyrand, in which the latter proposed a complete plan for the partition of Belgium, acceptable, he thought, even to England. The Belgians were unworthy of independence, while moreover there was no solution in sight to terminate the French occupation; their withdrawal without tangible results, such as the demolition of fortresses on the frontier of France, would mean the end of Périer's Cabinet. Prolonged occupation would bring about the fall of Lord Grey's administration Palmerston was surprised, but his information supported Bülow's: he had just learned that Marshal Gérard had received orders to hold the Belgian fortresses until he should receive clear contrary instructions from the Minister of War, and the newspapers wrote "that the French Govern-

[21] Binder to Metternich, 2 Aug., Gourieff to Nesselrode, 6 Aug., Waldburg-Truchsess to Frederick William III, 9 Aug., Ged., III, 211, 490, 44; Bagot to Palmerston, 9 Aug., Bagot MSS; Frederick William III to Empress Nicholas, 16 Aug., *Aus dem letzten Jahrzehnte Friedrich Wilhelm III,* p. 158.

[22] Talleyrand to Sebastiani, 11 Aug., A.é. Angleterre; Falck and Van Zuylen to Verstolk, 11 Aug., R.A.; A.R. Falck, *Gedenkschriften,* pp. 312-13; Mme Lieven to Alex. Benckendorff, 12 Aug., Lie MSS.

ment is collecting another corps as a reserve for Gérard." "Bülow's notion is", wrote Palmerston to Grey, "that this move of the Dutch had been planned between Talleyrand and Falck, who have been much together..... [lately] and that Russia had not discouraged it, partly wishing to help Holland to a share of Belgium and partly thinking that war in the West would leave her elbow room in the East." Palmerston may also have had in mind Bagot's report of 2nd August, in which the Ambassador said that Van Zuylen, before he left for London with his eventful notification, had only communicated with the French *Chargé d'affaires.* The Foreign Secretary assured Bülow that the British Government would find a way to get the French out of Belgium; Bülow thought Palmerston relieved by Granville's dispatch received on the morning of the 13th, but on the same day an angry letter was sent from Stanhope Street to The Hague as it appeared that fighting still continued in the Low Countries and that this offered an opportunity for Gérard's forces to advance further into Belgium.[23]

Grey had already become "a little anxious" with regard to French promises of evacuation, as the Paris Government tried to introduce other motives than the Dutch advance for the French occupation of Belgium, and reports from the British Ambassador could not but strengthen his and Palmerston's anxiety. Sebastiani and Périer immediately agreed that the French troops should avoid approaching Antwerp, Venlo and Maastricht – as stipulated by Protocol No. 31 – but the former raised many objections to the occupation of Maastricht and Venlo by Prussian troops, which was then being contemplated in London; he even suggested that Holland should lose those places to Belgium as a punishment for the breaking-off of the suspension of hostilities. He even tried to make the British Ambassador agree to the prolongation of the stay of the French army in Belgium until the border fortresses were demolished or at least until some agreement on this subject was reached. On the

[23] Palmerston to Grey, 12 Aug., Webster, II, 818; Bülow to Frederick William III, 13 Aug., Franqué, p. 313; Palmerston to Bagot, 13 Aug., Bagot MSS. Only the day before his conversation with Bülow Talleyrand wrote the following (11 Aug.): "L'Angleterre devait bien trouver dans tout ce qui vient de se passer en Belgique des motifs pour croire qu'il n'y a pas de Belgique possible et que c'est par des idées de partage que l'Europe trouverait la garantie positive d'une paix générale. Mais l'Angleterre est bien éloignée de cette idée." Talleyrand, *Memoirs,* IV, 260. It will be seen that, within a few days, he arrived at a different conclusion. Cf below, p. 205.

14th came the bombshell: Marshal Soult, the French Minister for War, told his Parliament that the Dutch were leaving Belgium but "that the French troops would not on that account return into France". Only with difficulty did Granville succeed in making Sebastiani declare in the Chamber that the Government hoped also to be able to withdraw French forces soon.[24]

This caused no little surprise in London but if the real purpose of the French Government was the enforcing of the demolition of the fortresses, there were other reasons also, equally well founded and – under different circumstances – likely to be appreciated by the British Government as well. In a special interview Granville obtained a very detailed statement on the subject from Louis Philippe himself, namely "that the spirit of the promise made by the French Government was that their troops should evacuate Belgium when it was secured against another sudden attack from Holland, for unless an engagement was entered into by the King of the Netherlands not to recommence war, Belgium might be again overrun, & the Prince of Orange proclaimed King at Brussels before the French Army (if situated within its own frontiers) could reach that city and even the Belgick fortresses be in danger of being occupied by Dutch troops." [25]

At this stage, however, the Ambassador did not admit the French reasons, though he soon came to appreciate Louis Philippe's predicament. From London, however, before even Soult's statement became known, Talleyrand implored his Government to withdraw its troops immediately. Van Zuylen was now delighted to report the reserved attitude of the Eastern diplomats in London: it was only too obvious that it was the British Cabinet which was compelled to seek their support, especially in view of the assurance given to Parliament of a speedy evacuation of Belgium by the French. Even Lord Holland was somewhat shaken in his French sympathies. The action of the French Government was a powerful blow both to the cause of peace and to the Reform Cabinet "The French must disarm this hostility to us...." he wrote to Granville, and he concluded: "If the French wish us to concert with

[24] Grey to Granville, 13 Aug., Howick MSS; Granville to Palmerston, 14, 15 Aug., F.O. France. Périer excused himself from going into the matter, while Sebastiani, until the "Moniteur" was shown to him, maintained that the Press had given an incorrect account of Soult's speech.

[25] Granville to Grey, 15 Aug., Howick MSS.

them now or hereafter about the Poles they should take care not to alienate the public mind here by conduct or language in Belgium." [26]

Fortunately for France there was no official notification of the Dutch retreat and the Conference countersigned by their Protocol of 18th August the prolongation of the French occupation. Talleyrand was only too glad for this help in continuing his "idée favorite". Next day, in spite of his earlier warnings, he wrote to Paris: "La force des choses amènera indubitablement un partage de la Belgique, l'Angleterre elle-même sera forcée d'y consentir." The idea seemed to have gained currency in London. It was acceptable at the Russian Embassy. Van Zuylen believed that all the friends of Holland admitted the impossibility of Belgium continuing as an independent state, but that Grey was invincibly opposed to any new solutions because he was a most convinced opponent of the partition of Poland 40 years before.[27]

In Paris Granville had no more success. Louis Philippe's confidential communication was made public by the King's statement in Parliament that he would order the retreat of this troops from Belgium as soon as he obtained a guarantee "que le retour de notre armée n'exposera pas la Belgique à de nouvelles agressions." This necessarily caused Granville to protest and Sebastiani professed his regret at the lack of confidence in the French Government: the French would withdraw but they had a right to require security for Belgium. On 22nd August the Minister declared formally that the withdrawal was independent of any other consideration, but he could not make this declaration, he said, "without at the same time stating that the French Government claims the right to negotiate with the Belgic Government respecting the dismantling of the fortresses." This in his turn the Ambassador was not prepared to accept – the matter could only be discussed between Belgium and the Powers responsible for the construction of the fortresses. The French moderated their tone, and, in all frankness, both Sebastiani and Périer "earnestly begged me to consider the critical position of the French Government in regard to the fortresses, the demolition of which was pronounced in the King's speech." [28]

[26] Talleyrand to Sebastiani, 15 Aug., A.é. Angleterre; Van Zuylen to Verstolk, 16 Aug., R.A.; Holland to Granville, 16 Aug., Granv. MSS.

[27] Talleyrand to Sebastiani, 19 Aug., *Revue des Deux Mondes,* LVI (1910), 333; Mme Lieven to Nesselrode, 22 Aug., Lie MSS; Van Zuylen to Verstolk, 20 Aug., R.A.

[28] Granville to Palmerston, 18, 21, 22 Aug., F.O. France.

The second condition of the French, a special pledge from the King of the Netherlands not to reopen hostilities, was even less likely to be obtained. When the evacuation of the Dutch troops was known in London, feeling towards The Hague very much improved according to Falck and Van Zuylen; the Austrian members of the Conference also indicated sympathy for the Dutch cause. There were speeches in favour of Holland in the House of Lords. There was even an article in "The Times" in which the action of the Dutch troops was described as the result of the Conference's protocols. France could never, for obvious reasons, act as the Conference's agent: it was the French influence which brought about the substitution of the "Bases de séparation" by the XVIII Articles. Louis Philippe thought it necessary to have this article refuted by the French Press. The diplomats of course felt differently about the Dutch move. Bagot did not agree with William I's methods, but, much as he realized that these accounted very largely for the insistence of the French to remain in Belgium, he nevertheless doubted whether the King could be induced to relinquish his freedom of movement with regard to Belgium. "In marching into Belgium," he wrote to his colleague at Brussels, "the King committed an act of most mischievous imprudence. In marching *as he did* he was guilty of very unwarrantable duplicity. But he has *succeeded,* and however that success may anger the Conference, it has produced as success always will do, a great moral effect in his favour; – and the consequence now of any attempt, on the part of the Five Powers, to punish him for it, will be not to make him more tractable, but to engage the sympathies of mankind on his side. I hope that the Conference may feel this." [29]

Indeed, this was very much the case. Though Palmerston wrote to Berlin that King William I lost all consideration he might have expected, the Protocol of 23rd August (No. 34) imposed no hardships upon him, the conditions being suspension of hostilities for 6 weeks under the guarantee of the Five Powers, and the line of demarcation to be that which existed before 2nd August. Palmerston lacked the support of the other members of the Conference; he believed moreover that Prussia was not against a partition of Belgium, and

[29] Falck and Van Zuylen to Verstolk, 20 Aug., R.A.; Esterhazy and Wessenberg to Metternich, 20 Aug., Arneth, *Johann Freiherr von Wessenberg,* II, 123-4; Louis Philippe to Sebastiani, 21 Aug., *Revue des Deux Mondes,* (1910) LVI, 318; Bagot to Adair, 20 Aug., Bagot MSS.

Russia would welcome any new troubles in the West of Europe.[30] Protocol No. 34 was agreed upon on the receipt of the Dutch notification that their evacuation would be completed by 21st August, and yet not a single word was said in it about the French troops in Belgium. And yet Granville heartily sympathized with the precarious situation of the French Government: if they subscribe to the Protocol, he thought, they had better resign, for "neither the Chambers nor the public would tolerate the withdrawal of the French troops from Belgium with no other security... than such as was to be derived from a mere promise of the King of Holland." French protests followed, as shall be seen. In the meantime, the first battle for the withdrawal of the French was lost.[31]

Moreover, Palmerston could not be satisfied with King Leopold, who, to England's surprise, himself requested the prolongation of French occupation. In the flow of arguments urging the King to adopt British policy for his own, the Foreign Secretary did not forget to add that before long the Polish contest would finish and allow Russia and her allies more freedom in supporting WilliamI.[32]

But King Leopold was bitterly disappointed with the Conference's lenient attitude towards the King of the Netherlands. He had already been imprudent enough to refer to the question of fortresses through Lebeau's letter of 2nd August, requesting French assistance, but "the language in which the Conference was pleased to remonstrate with the King of Holland..... was certainly not calculated to stop an army marching upon Belgium" and therefore he tried to retain at least his one convinced supporter. For the rest, the French were only too eager to fulfil this rôle. Adair was convinced that the French entry took place *before* Marshal Gérard obtained Leopold's invitation. Notwithstanding

[30] The Protocol in B.F.S.P., XVIII, 830; Palmerston to Chad, 19 Aug., Palm MSS; Palmerston to Granville, 23 Aug.: "The French wish to stay in; the Prussians do not know their own mind on the subject because they always have a secret thought that if the French stay, and war ensues, the partition must follow and they will come in for their share, and hence it is that Werther is not very keen on the subject though Ancillon protests to Chad that his only wish is to get the French out. Austria is the nearest to us... Russia who, I suspect, knew more of the Dutch King's inroad than she would choose to own, ... would not be sorry to see us quarrelling together." Bulwer, II, 121.

[31] Granville to Palmerston, 26 Aug., F.O. France.

[32] H.C.F. Bell, *Palmerston,* 1936, I, 134; cf. Palmerston to Grey, 22 Aug., Howick MSS.

the presence of an Ambassador, another special envoy, Count Latour-Maubourg arrived in Brussels on 19th August to reach an agreement on the fortress question before the departure of French troops.[33]

In London, for Wellington as well as for Grey this presented a virtual *casus belli.* To the slight surprise of Mme Lieven, Grey ("qui est toujours singulièrement curieux de connaître les opinions du duc de Wellington et se règle constamment sur elles") maintained that England was prepared for a war, as she had lately been increasing her armed forces, only with less publicity than had the others. Nevertheless, Grey would have preferred diplomacy but realized that for Talleyrand all his arguments were placed in an unfavourable light by the incapacity shown by the Belgian Government. "La combinaison Léopold pesa sur lui de tout son poids", observed Mme Lieven. Accounts of the Prince of Orange's successes, even the conduct of the Duke of Saxe Weimar with regard to Lord William Russell, all this completed the report of the failure of British policy sent to St. Petersburg: "Tout cela est fort humiliant pour l'Angleterre... a mis dans toute son évidence la pauvreté des moyens de ces braves Belges aux gasconnades desquels on a fait tant de sacrifices, et enfin.... a amené l'armée française en Belgique." In these circumstances "Lord Grey me demande une bataille ou la prise de Varsovie avec autant d'impatience que j'en éprouve moi-même," noted Mme Lieven. From London, however, Russian prospects were by no means clear, even to her.[34]

Surprisingly enough, two days later Grey informed Mme Lieven that he could not refuse to see another Polish envoy, the poet Niemcewicz, who had arrived in London on 21st August, bringing a letter from the Polish Government to the King. A survey of Polish affairs during August will be helpful to clear this apparent discrepancy.[35]

[33] Martinet, p. 105, cf. above, p. 239; Leopold to Grey, 16 Aug., Howick MSS; Adair to Palmerston, 11 Aug.; "The proof is most clear that the French entered those territories not only without leave but against the express will of the King", as Lord William Russell found them at Mons at 2 o'clock on the day on which, about noon, the King had written to Gérard. Palm MSS. Neither Adair nor obviously Grey could have known that the French entry was first authorized by the Belgian Cabinet, cf above, p. 200.

[34] Mme Lieven to Nesselrode, 22 Aug., closing as follows: "Paskévitch, est-il donc attaqué et paralysé comme l'était Diebitsch?" Lie MSS.

[35] Mme Lieven to Grey, Grey to Mme Lieven, 26 Aug., *The Lieven-Grey*

III

Before the Dutch campaign Palmerston informed the Ambassador in St. Petersburg that the reason for refusing the French offer concerning Poland was his conviction that after a new battle which now seemed imminent, Nicholas might be more inclined to treat. The Eastern Courts, however, remained unconvinced about the likelihood of indifference on the part of Britain and France in Polish affairs. Metternich hoped that the Polish movement might collapse before "revolutionary" public opinion in England and in particular the fall of Périer's Cabinet would compel the two Powers to intervene in Poland. Militarily, he hoped, Paskevitch should be able to secure victory during August. Berlin seems to have been less confident about Russia's success; there one could more readily picture the military situation from first-hand accounts. Bernstorff would be glad if France advised the Poles to submit, for then, he told Flahaut, he had grounds to believe that Nicholas would meet them half-way.[36]

In addition the Russian Government, having formally refused Mortemart's overtures, was by no means easy about further developments of French policy with regard to Poland. Louis Philippe's speech to Parliament, the publication of which was suppressed in Russia, gave rise in government circles to the comment: "Le Roi des Français a jeté le masque", but the Dutch commentator, unlike Mortemart, saw no danger for Russia in the King's words relating to Poland, except in so far as they gave evidence of the abandonment

---

Correspondence, II, 269-271. Grey wrote: "After having seen Walewski, after both he and Wielopolski had been *frequently* [italics mine] received by Palmerston I thought I could not very well decline an interview which was proposed to me by a person whom I had formerly known when he was here with Kosciusko." Niemcewicz was an A.D.C. to Kościuszko, the American Independence War hero, during Kościuszko's unsuccessful campaign against Russia and Prussia in 1794 (followed by the third partition of Poland), whereafter, after a short captivity in Russia, he came with Kościuszko to England and then went to America.

[36] Palmerston to Heytesbury, 2 Aug., Heytesbury Papers; Metternich to Wrede, 2 Aug., V. Bibl, *Metternich in neuer Beleuchtung*, 1928, p. 236; cf. above, p. 192; Chad to Palmerston, 3 Aug., F.O. and priv., Palm MSS. Chad was then instructed to protest again against the breach of neutrality, and intended to submit a formal note; he later refrained from doing so in the belief that France wanted to break the alliance of the Four Powers "& especially detach Prussia from England & perhaps to make use of the sympathy felt here for the King of Holland for that purpose." Palm MSS.

of Poland and might in consequence drive the Poles to desperate intransigence.[37]

Meanwhile the French Chambers seemed to have estimated the King's speech above its face value, as far as it concerned Poland. In the long debate on the address in reply, held after the news of Dutch invasion and Franch intervention, the motion of Lafayette for the recognition of Poland was defeated by a narrow majority. Another motion proposed that the Chamber should see in the King's words the certainty (certitude) that Polish nationality would not perish, but eventually the weaker expression "l'assurance" was adopted. Périer, according to Granville, was prepared to resign if these motions were carried; a special committee was appointed to revise the passage in question. Yet the French Premier remained vague about his intention of remaining in office and the British Ambassador was troubled by these discussions.[38]

The Polish envoys in Paris were seen by Sebastiani at the beginning of August. He seemed to have known nothing about the English *démarches* in Berlin with regard to Prussian neutrality, and he was glad to hear of them. As for himself he only expressed his hope that the Poles would hold out during this campaign; Kniaziewicz replied that this could be done, if the French Government made a straightforward declaration in favour of the Poles. This remark met with no response.[39]

The two envoys hardly realized the effect of the hostilities in the Low Countries on their cause. On the 8th, rightly, as it seemed, the Chamber being engaged in discussion of the "mention polonaise" in the address in reply, they approached Sebastiani again, but he refused to do anything concerning Poland at that time. Writing to Walewski, they were wondering if England would allow the French to occupy Belgium.[40]

[37] Extracts of letters from St. Petersburg, dated 6 and 10 Aug., in Van Zuylen Papers, R.A. The second letter reads: "Les phrases sonores qui concernent la Pologne ne sont autre chose qu'une ironie sanglante, elles prouvent que la médiation offerte par la France a été rejetée . . ."

[38] J. Dutkiewicz, *Poland and France in 1831*, pp. 128-9; Granville to Palmerston, 8 Aug., F.O. France. The Netherlands Minister in Paris believed that an agreement between the Government and the Parliament had been reached. Fagel to Verstolk, 8 Aug., R.A.

[39] The Paris envoys to the Polish Government, 1, 9 Aug., to Walewski, 10 Aug., B.P.

[40] Walewski to the Paris envoys, 5, 7, 12, 17 Aug., B.P. The last letter reads: "Tout le monde est très occupé du discours de Maréchal Soult dans lequel il dit

Walewski had a clearer view of the situation. Already on the news of the resignation of Périer he experienced difficulty in getting anybody to listen to him talking about Poland, whereas in his interview with Palmerston on the 7th he found the Foreign Secretary cold and occupied with the expectation of war because of the development in Belgian affairs. Palmerston again spoke about his *démarches* to Prussia, but Walewski knew from his correspondence with Flahaut that there was little hope of success in Berlin. "L'Angleterre ne fera rien", wrote the Polish envoy on the 12th. The same report refers to the first, very short, debate on Poland in the House of Commons. One of the four Members who took part in it, Mr. Hunt, informed the House that a memorandum, presented by the Westminster Union, calling the Government's attention to the war in Poland had not even been answered by the Foreign Secretary. Mr. Hume added that "he had met no individual" who was not interested in the measures that the Government intended to adopt with regard to Poland. Col. Evans reminded the House of his intention to bring forward a motion concerning Poland. Following the speech of Sir Robert Inglis, not even a vote was taken on the printing of the memorandum. "Mais du moins Lord Palmerston n'a rien répondu de défavorable" commented the Polish envoy after stating that little attention had been paid to the speeches.[40]

More, of course, had been expected of Sebastiani and the Paris mission was greatly disappointed. They sent a memorandum concerning Prussia to the French Prime Minister and hoped for the best as three days later they received an acknowledgment, the first such during their activity in the French capital. Then, in their 25th interview with Sebastiani every hope was dissipated. The French Minister not only repudiated their suggestion to connect Polish and Belgian affairs, in the same way as Talleyrand had proposed this some days earlier – but was not even prepared to admit any breach of neutrality by Prussia. He left unanswered questions concerning his recommendation, made at the time of the expected Anglo-French *démarche* to Russia, to hold out for another two months so as to let negotations mature; on the contrary, he advised the Poles to treat with Paskevitch. He did not, however, fail to communicate to Granville Mortemart's

que l'armée française ne se retirera pas de Belgique. C'est entièrement pour l'Angleterre un casus belli."; *Hansard Parl Debates,* V (3rd Series), p. 950. Cf. J. H. Gleason, *The Origin of Russophobia in Great Britain,* 1950, p. 121.

dispatch of 23rd July, expressing his hope that it might be considered by the British Cabinet.[41]

It may be worth while mentioning at this stage that the failure of the Polish agents, even before it had become so obvious, was subject to criticism on the Polish side as well. At the end of July Count Jelski, a former director of the Bank of Poland, arrived in London. He had been interned by the Austrian authorities and had recently escaped from Bohemia. Jelski entirely disapproved of the action of the Polish envoys in Paris and of Walewski as regarded the XVIII Articles, believing that the rejection of those proposals by the Belgians would bring England closer to France. It has been shown already that this reasoning had also been that of Palmerston up to a certain point, and that in any case Palmerston tried to impress this possibility upon the Eastern members of the Conference. Jelski developed his views to Lord Holland. He suggested that the "inactive" attitude of Prussia towards the Polish revolution, announced to Chad, was a counter measure of the Berlin Government, showing their concurrence with the King of the Netherlands in his rejection of the XVIII Articles. Lord Holland saw in it only a mark of deference to Nicholas; yet he believed that Prussia would not vote for war, if she were told by Britain that she only knew neutrality or war. The British Government was already working in this direction, added Lord Holland. The French mediation proposal had to be rejected, he continued, because it was not clear whether France meant war or not. Jelski seized this opportunity to say that the best way to avoid war would be to request the King of Prussia to offer mediation – to such an offer Nicholas might be likely to subscribe; in the meantime, delay could not but complicate matters, "puisque la France ne pourra s'empêcher de nous reconnaître au premier avantage remporté par nos troupes et cette démarche pourra amener une guerre qu'on évite avec tant de soins." Lord Holland seemed to agree entirely; he requested Jelski to repeat this reasoning to Palmerston and promised to obtain for him an interview for this purpose. Moreover, he said that the recognition of the Polish Government by Britain seemed likely to him. "C'est un gouverne-

[41] Périer to Kniaziewicz, 13 Aug., the Paris envoys to the Polish Government, 16 Aug., B.P. cf. above, p. 199; Granville to Palmerston, 18 Aug.: "...if you allow him [Nicholas] to do as he pleases Poland is annihilated for nothing can exceed the hatred against them [the Poles]". This is almost a translation of Mortemart's dispatch, referred to on p. 190. F.O. France.

ment de fait qui existe depuis huit mois, commande aux troupes, livre des batailles et remporte des victoires. Un gouvernement comme celui-là peut être reconnu à tout moment." Yet he was not prepared to bring the matter before the Cabinet. "Je doute pourtant", he added, "que nous puissions nous refuser à le faire aussitôt que les événemens qui paraissent prochains auront décidé d'avantage les affaires en votre faveur." This condition made a great difference to the Poles, who, mindful of the advice of Sebastiani, avoided a major battle lest it should be unsuccessful. But more important than that, the news of the war in the Low Countries was on its way to London.[42] Like Walewski, Jelski did appreciate the effect of this event on international relations, but he failed to draw the appropriate conclusions. He thought that the general war which now appeared inevitable, would unite Britain and France on one side, in spite of London's fear of France, because Prussia would support Holland. Continuing his own reasoning Jelski now advised the Paris envoys that if an arrangement concerning Belgium should be reached without reference to Poland, the Polish cause would be lost. In a long letter to Talleyrand Jelski expounded the necessity for making the neutrality of Prussia in the Polish war the condition of peace. The Ambassador sent his letter to Paris and in his own dispatch at that time advocated such a possibility even without actual war. Sebastiani, as has been shown, did not accept Talleyrand's proposal, and before long it became clear that France had to keep her army in Belgium at all costs and could not make any conditions, the occupation itself being hardly tolerated by the other Powers.[43]

On 16th August Jelski heard from Lansdowne that the Cabinet was discussing Polish affairs. The news from Poland, neither reliable nor wide-spread did not contain anything of importance. Chad reported that there was dissatisfaction with the inactivity of the armed forces and little confidence in the Commander-in-Chief,

[42] Jelski to Czartoryski, 28 July, 4 Aug., B.P. For Palmerston's interview with Bülow in June, cf. above, p. 159.
Although the majority of the Polish generals in the council of war on 27 July voted for a battle, the C.-i-C., Skrzynecki, on hearing of Sebastiani's statement of 8 July, reverted to his former policy of inactivity. Barzykowski, *History of the November Rising,* IV, 372; cf W. Tokarz, *Wojna Polsko-Rosyjska w r. 1831* (The Polish-Russian War in 1831), 1930, p. 484.

[43] Jelski to Czartoryski, 4 Aug., to the Paris envoys, 5, 6, 8 Aug., to Talleyrand, 7 Aug., B.P.; for Talleyrand's dispatch see above, p. 199.

Skrzynecki. But soon afterwards the news spread in London of the arrival in Warsaw of the corps of General Dembiński from Lithuania which, it was believed, meant a breathing-space for the Poles. In fact, though it could be considered a military success in itself, it nevertheless meant a complete failure of the rising in Lithuania, where there were now no regular forces left, and Dembiński's corps, composed of newly recruited soldiers, only represented a very modest contribution to the strength of the Polish army.[44]

A few days later Niemcewicz, a very distinguished envoy, then in his seventies, arrived in London. The object of his mission was to take part in negotiations concerning Poland which the Polish Government thought, after obtaining information of the French *démarche* of 7th July, were about to take place in the British capital. Nothing of the kind was contemplated in London when he arrived after a journey of more than three weeks and all that could be undertaken was to try to present the letter from Prince Adam Czartoryski who, as president of the Polish Government, had written to the King.[45]

The application from the Poles to present Czartoryski's letter coincided with the arrival of Chad's dispatch containing information about new disturbances in Lithuania, east of the provinces so far affected, and about the corps of General Rüdiger which was supposed to march upon Warsaw from the south, but had not crossed the Vistula at all.[46]

In a short note, dated 24th August, Palmerston informed Grey of the Polish application, and of his intention to refuse the letter; he accordingly gave Niemcewicz no hope of the possibility of his being received by the King. But the mere news of Niemcewicz's arrival greatly disturbed the wife of the Russian Ambassador. She had already done her best to persuade Grey not to see the new Polish envoy; but on learning the envoy's name, she appeared alarmed: "...j'avoue que souvent lord Grey m'a parlé de lui comme d'une très vieille et très intime connaissance..., cela

[44] Jelski to the Paris envoys, 16 Aug., B.P.; Chad to Palmerston, 7 Aug., F.O. Prussia; Van Zuylen to Verstolk, 20 Aug., R.A.

[45] The Polish Government to the Paris envoys, 14 July, Walewski to the Paris envoys, 25 Aug., B.P.

[46] Grey to Mme Lieven, 26 Aug., *The Lieven-Grey Correspondence,* II, 271. Only the second report was true. The same letter reports the arrival of Niemcewicz, cf above, p. 208.

justifie la visite [at Grey's] mais combien je suis contrariée de ce que ce personnage soit ici! Il est connu personnellement de plusieurs ministres; et entre lui et Talleyrand ils feront la bien mauvaise besogne, si Paskévitch ne se hâte pas."

Her very long letter on this subject is only the preamble to the statement that, short of Paskevitch's immediate success, Europe is lost; this is supplemented by a remark on Grey's complaints about Leopold, Wellington's view that only one of the Belgian fortresses could be dispensed with, and Mme Lieven's own belief that the French would not leave Belgium.[47]

About that time the Polish envoy in Brussels, Załuski, having obtained a favourable reception on the part of the King, asked the Minister for Foreign Affairs whether the King of the Belgians could not be requested to write to the King of England in favour of the Poles. The Belgian Minister apparently avoided this embarrassing question, but told him in confidence that the British Ambassador had begged him not to make any difficulties – a clear reference to Leopold's request for the maintenance of a French force – as the British Government was only awaiting a settlement in Belgium to be able to intervene in Poland. These difficulties, the Minister thought, had ceased to exist, as on Leopold's request the French troops were leaving at once, "et il ne restera qu'un très petit corps sur la demande du Roi lui-même, et qui sera à sa disposition." [48]

The Polish envoy did not stop to consider the implications of this final remark, but immediately went to see Sir Robert Adair. The reception was friendly and warm beyond Załuski's expectation, and in the course of a long conservation, in wich the Ambassador gave evidence of his Polish sympathies, the Polish envoy obtained the following statement: "Vous avez raison de désirer que la question Belge soit terminée, car enfin cela nous pourrait mettre dans le cas de nous occuper activement de la vôtre. Je ne vous dis cela comme une certitude, mais je l'espère et cette espérance n'ait qu'un sentiment vague mais fondé sur les données." To this the Polish envoy replied that English intervention had been looked for long

[47] Palmerston to Grey, 24 Aug.,: "The Polish Government at Warsaw sent a Deputation to London with a letter to the King to ask the assistance of England. Walewsky incloses a copy of this letter. I conclude that I must tell him that under existing circumstances the letter cannot be presented to the King." Howick MSS; A. Czartoryski, *Żywot J. U. Niemcewicza* (The Life of J. U. Niemcewicz), 1863, pp. 363 sq.; Mme Lieven to Nesselrode, 25 Aug., Lie MSS.

[48] Załuski to the Paris envoys, 23 Aug., B.P.

enough, but Adair repeated that the Belgian settlement must be achieved first, and added that "l'intérêt que porte tout le peuple Anglais à votre héroique Nation est partagé par ceux qui sont à la tête du gouvernement; mais depuis que la politique de Mr. Pitt et tous ses successeurs, en nous ayant forcé pendant 13 ans à faire une guerre injuste à la France et aux principes de la liberté, nous a grevés dans une dette publique immense, nous sommes obligés à une grande circonspection" – otherwise English intervention would not have had to be awaited for such a long time. Adair went on to say that he strongly advised the Poles to try to enter into some negotiations with Austria. From this personal knowledge of the Emperor Francis he was sure that the Emperor was favourably disposed toward Poland; moreover, even now, before leaving London the Ambassador had, during a conversation with Esterhazy, become convinced that Austria would lend a willing ear to Polish propositions.[49]

It will be noted that, although unconnected with Niemcewicz's visit to Grey, of which at this time neither Załuski nor probably Adair could have known, Adair's statement concided with the Cabinet meeting devoted to the Polish letter and had only been preceded by a day or two by the arrival in London of Leopold's request for permission to keep a French force in Belgium; this request, warmly supported by Talleyrand, constituted for the French Government the only hope of a settlement on the subject with Britain.[50]

The meeting took place on 25th August. There are, in the Broadlands archives of Palmerston, the written opinions of eleven Ministers, nine of whom were members of the Cabinet, on whether the letter from Prince Czartoryski could be presented to the King. None of the Ministers was in favour of allowing a delegation to present it. Although political motives which led some of the Ministers to advise the acceptance of the letter are hardly to be found a survey of their opinions provides some reflection of the feeling towards the Polish movement. The list begins with Palmerston who took the formal view: the letter is signed by the President of a government not recognized by Britain, "which came in existence through revolt against an allied sovereign." Grey inclined, with some reluctance, to Palmerston's opinion; he wrote, however:

[49] ibidem.
[50] Palmerston to Grey, 22 Aug., Howick MSS; cf. below, p. 219.

"Might not the answer be that the letter had been delivered but that under the present circumstances H.M. could not receive the deputies or authorize his Governement to enter into any further communication about it? But I really should like to hear what others have to say before this question is decided," Lord Melbourne, Sir James Graham, Mr. Stanley and Lord Goderich anwered decidedly: no. Lord John Russell added that the presentation of the letter would only lead to false interpretations. "But at the same time the British Government was willing to admit in the clearest terms that the signature of the Prince Czartoryski would under other circumstances be entitled to the utmost regard." The remaining four Ministers were in favour of presenting the letter. Lords Lansdowne and Carlisle from motives of deference towards the Poles, to which Mr. Grant added: "It might be humble request for mediation and give opening to peace." Lord Holland alone made a more extensive statement. He said that he would be in favour of allowing the deputation though aware that this implied the recognition of the Polish Government. "I think that this question [is] well deserving the early consideration of the Cabinet." Before, however, this was decided upon, the delegation could not be admitted to the King but the letter might be received without scruples.[51]

The majority being against its acceptance, the letter was refused, though on the request of Walewski Palmerston promised to reconsider this decision. For the rest, he was prepared to act immediately if Warsaw fell. He wrote unofficially to the missions in St. Petersburg and Berlin. "We are anxious for accounts from Poland" he told Lord Heytesbury. "The next news must acquaint us with decisive events." The lines immediately following reveal Palmerston's opinion about the state of the Polish movement. "It is a great pity that the Emperor's proclamations are not calculated for Poland instead of Petersburgh. To call upon an armed nation to submit at discretion upon a mere promise in a general form of clemency and pardon, may serve some purpose, but cannot possibly have the slightest effect upon the Poles; nor could any man in his senses do what those proclamations require. If the Emperor stated what he would do when the Poles should have submitted and engaged to secure to them the constitution and privileges, the violation of

[51] File 3, Cabinet Ministers, Palm MSS. See Appendix XII. Cf. Webster, I, 187, where a slightly different picture is given.

which was the cause of revolt, then indeed a promise of pardon.... might tend to some good." It was evident , however, that Nicholas wanted to abolish the constitution. Should the Poles be defeated, the Powers who were parties to the Treaty of Vienna must watch the situation to prevent this abolition. The Minister in Berlin was again informed that the British Government knew of the assistance in every form given by Prussia to the Russian army; Chad was therefore instructed to explain to the Berlin Government that this activity would encourage other governments to help the Poles. Palmerston added also complaints about British passports not having been respected and passage of goods disallowed – this again had reference to Poland.[52]

However, there was no difference in the attitude towards the Polish envoys in London, and their report to Paris, dated 29th August, indicated that they acquiesced in their failure to obtain some favourable action on the part of the British Government. They had been seen by Palmerston, and Niemcewicz also obtained interviews with Grey and with Lansdowne. "En résumé, cela revient toujours à la même chose" they wrote. "Si nous sommes victorieux, si cette campagne finit à notre avantage, on ne laissera pas recommencer une seconde; si nous sommes vaincus, on interviendra en notre faveur et on demandera l'exécution des traités de Vienne, mais avant l'un ou l'autre de ces événemens on ne fera rien." The letter to the King would not be accepted and no more would be done with regard to the neutrality of Prussia. As it happened, at this time another report on the subject was on its way from Berlin.[53]

Only a day after Walewski and Niemcewicz had written their report unhappy news from Warsaw spread in London. On 15th and 16th August riots had taken place in Warsaw, accompanied by atrocities against Russian internees and Polish prisoners suspected of treason. These events had a disastrous effect upon the Polish cause in England. Van Zuylen and Greville, as well as Mme Lieven and Countess Nesselrode, all stated that the Poles thus lost most of the goodwill prevailing in London with regard to their

[52] Walewski to the Paris envoys, 26 Aug., B.P.; Palmerston to Granville, 26 Aug., Webster, I, 188, to Heytesbury, 26 Aug., Heytesbury Papers, to Chad, 26 Aug., Palm MSS.

[53] Walewski and Niemcewicz to the Paris envoys, 29 Aug., B.P.; Chad to Palmerston, 22 Aug.: Ancillon said that only diplomatic messengers and doctors were allowed into Poland. Palm MSS.

struggle; at the Russian Embassy considerable political consequences were attributed to the news of the Warsaw riots. As much could be said on the other side. Grey found Niemcewicz in total despair, and believed that "this unhappy business" must already be settled by this time. News of a different nature was yet to come, but the general view that at the end of August the revolt was about to destroy itself was apparently established.[54]

IV

In the meantime the question of the evacuation of the French from Belgium dragged on. The route from Calais to Paris might have been covered by couriers with instructions to Granville "de ne laisser à M. Casimir Périer trêve ni repos" until he should have issued the order of withdrawal to the French troops – as Palmerston colourfully put it to the Dutch – but this obviously did not amount to much; moreover Leopold considered himself obliged to request the assistance of a French force. At this time Grey seemed apologetic in his letter to the Belgian King, whilst professing his readiness to help him against the Dutch invasion – had he known of it; it was only because "a fleet not being so easily effected or put into action as *an army already in state of preparation on the frontier of the wounded country*" that there was no help from England. The more merciless for King Leopold must have been Grey's stress on a speedy evacuation of Belgium by the French forces.[55]

Indeed, apart from all other considerations, the French fears with regard to the security of Belgium were justified. To Granville the parliamentary difficulties for the French Cabinet regarding the question of evacuation were already becoming obvious. Their anxiety was conveyed to Talleyrand in most alarming expressions. After

[54] *Greville Diary*, II, 197; Van Zuylen to Verstolk, 30 Aug., R.A.; Mme Nesselrode to her husband, 2 Sept.: "Comme cet événement a changé la face des affaires, disent-ils [Lieven and Matuszewic], quelle mine piteuse a Talleyrand, et combien le ministère d'ici tient un autre langage, depuis qu'il espère que, s'il y aura la guerre, nous pourrons envoyer notre contingent." Nesselrode, *Memoirs*, VII, 224; Mme Lieven to Nesselrode, 30 Aug.: "...comme cela va changer la face des affaires de l'Europe, cela et l'heureux épisode de la guerre des Hollandais... que je regarde comme le premier signe de retour à l'ordre." Lie MSS; Grey to Mme Lieven, 31 Aug., *The Lieven-Grey Correspondence*, II, 272.

[55] Falck and Van Zuylen to Verstolk, 23 Aug., R.A.; Grey to Leopold, 26 Aug., Howick MSS. Italics mine.

observing that the Dutch had an army of 100,000 men and the Belgians "n'ont rien, absolument rien à leur opposer", Louis Philippe continued: "En vérité, mon cher Prince, je dois vous dire avec toute la franchise...., que je ne comprends pas comment cette situation de la Belgique, comment celle de mon gouvernement et la mienne vous ont échappé à tel point que vous n'ayez fait nulle difficulté de signer ce singulier protocol" (No. 34). Sebastiani, just as much alarmed, went into details: "La Belgique est dans l'anarchie, son armée est dissoute; le roi Léopold ne peut réorganiser ni son armée ni l'administration publique, s'il n'est pas protégé par une force quelconque" [56]

Besides, was he supported by England? The whole diplomatic corps at The Hague, complained the French Minister, including Sir Charles Bagot, had congratulated the Princess of Orange on the victories of the Prince's army. Consequently, in a most friendly way, Sebastiani tried to impress upon the British Ambassador that under such circumstances King Leopold would have no solution but to abdicate, "if he shall find that his demands for a division of troops to remain during a short time for his protection be rejected by the Conference." This was serious enough; General Baudrand who, as aide-de-camp to the Duc d'Orléans, had been in Belgium during the recent campaign, was sent to London on a mission to explain the French views to the British Government.[57]

In London current diplomatic news equally reflected the gravitiy of the situation. Informing Greville of the purpose of Baudrand's mission, the Austrian Ambassador told him that war was becoming inevitable. "He considered that the Polish business was over", noted Greville, "at which he greatly rejoiced." Falck and Van Zuylen reported that the two ships, recently added to Sir Edward Codrington's squadron and sent to the Downs, had taken troops on board. "La grande confusion en Europe est à la veille de commencer", commented Mme Lieven. "Cependant, elle commence sous de bons auspices: l'Angleterre est dans nos rangs." [58]

From Whitehall things did not look so precarious. Baudrand's reasoning was rejected in the first instance by Grey, but Talleyrand

[56] Louis Philippe to Talleyrand, 27 Aug., Sebastiani to Talleyrand, 27 Aug., Talleyrand, *Memoirs,* IV, 276-7 and 280-1.

[57] Granville to Palmerston, 27 Aug., F.O. France.

[58] *Greville Diary,* 31 Aug., II, 191; Falck and Van Zuylen to Verstolk, 30 Aug., R.A.; Mme Lieven to Nesselrode, 1 Sept., Lie MSS.

was able to report that no particular stress had been laid upon the date of evacuation. Indeed, Grey would rather wait for Leopold's declaration. Talleyrand then adopted a less co-operative attitude: he hesitated to sign a minor protocol, ordering the cessation of military preparations on both sides at Antwerp. "But Warsaw had already begun to tell with Prussia", observed Palmerston, and Bülow, becoming aggressive, said that he would propose to the Conference "to make a formal demand for the immediate retreat of every French soldier from Belgium." Palmerston then referred to the Protocol of 18th August (No. 33), (which provided for the withdrawal of the French after the Dutch should have done so). "All this brought Talleyrand to his pen and ink." [59]

The tension increased in Paris. On 3rd September, when Sebastiani informed the British Ambassador of his intention to leave 12.000 men at the disposal of Leopold, Granville answered that the British Government was not prepared to support the King of the Belgians against his own people. On a subsequent occasion Sebastiani argued that the security of Belgium had not been adequately provided for by Protocol No. 34. To Talleyrand he wrote that in view of the situation in Poland, William I might again try to entangle Europe in a general war; the French army should, therefore, stay in Belgium. Eventually, on the 7th Sebastiani told Granville that the re-establishment of the Kingdom of the Netherlands, whether by a Dutch invasion or by "intrigues of Holland" could not be allowed by France. The Ambassador protested sharply against France thus breaking the principle of non-intervention, but was satisfied that France meant war if the new régime in Belgium should be altered. Soon after the diplomatic corps in Paris had begun to deposit very formal notes on the Minister's desk, urging the withdrawal of the French troops notwithstanding Leopold's request, Sebastiani and Périer fixed in an interview with Granville the date of 30th September for the completion of the evacuation of Belgium.[60]

The Conference took up the subject of the evacuation with a renewed insistence after Protocol No. 37 on 31st August had coun-

[59] Talleyrand to Louis Philippe, 2 Sept., Talleyrand, *Memoirs,* IV, 248; Palmerston to Grey, 1 Sept., Howick MSS; Palmerston to Granville, 2 Sept., Bulwer, II, 124-5.

[60] Granville to Palmerston, 3, 5, 7, 10 Sept., F.O. France; Sebastiani to Talleyrand, 7 Sept., A.é. Angleterre.

tersigned the suspension of hostilities for six weeks. To begin with, as Palmerston explained to Grey, the French betrayed themselves: they professed their intention to stay until Leopold should have reorganized his army, but Baudrand told the Foreign Secretary that this could not be achieved within six weeks. Moreover it had become obvious to Palmerston that the Belgians wanted to keep the French army not so much for their protection but as a means of upholding their own pretensions in the forthcoming negotiations; they even succeeded in making Adair agree to the dismantling of a fortress as a gesture towards the French Government. No wonder, therefore, that the Eastern Powers pressed for a formal Protocol on French evacuation. With considerable difficulty the Foreign Secretary made them agree to Talleyrand's request to accept a non-publishable memorandum by the French Government on their intention to evacuate Belgium as a basis for the representatives of the Four Powers in Paris, where, after all, the final decision had to be taken. But no more time was allowed to the French Cabinet than was necessary for the return of a courier.[61]

It was Grey who tried to extract first from Leopold himself the declaration that he could dispense with the French troops, and this was only fair since their prolonged stay in Belgium was implicitly tolerated by the Conference on account of his request. Grey's insistence, to an inconsiderable degree shared by Palmerston, was only partially successful: at the same time that Granville was able to report that the French had given way, Leopold announced to Grey his intention of writing to Louis Philippe with the request for a complete withdrawal of all French troops from Belgium. "It is better..... that we should be somewhat exposed to danger," commented the Belgian King, "than that the harmony between England and France should be endangered." [62]

Neither he, nor Talleyrand, nor the other parties in the Conference were aware of this intercourse between Grey and Palmerston, and the French Ambassador, though persisting in his "système de retraite lente," sent repeated warnings to Paris. On the 5th it was "Les embarras du Ministère Anglais, résultans de la présence des troupes françaises en Belgique, semblent augmenter chaque jour." Van Zuylen and Esterhazy believed that the British Govern-

[61] Palmerston to Grey, 3 Sept. (three letters), 5 Sept., Howick MSS.

[62] Palmerston to Grey, 2 Sept. (two letters), Grey to Adair, 7 Sept., Leopold to Grey, 10 Sept., Howick MSS.

ment had abandoned the idea of keeping Périer in power as he seemed unable to control public opinion in France with regard to Belgium. On the 9th Talleyrand reported that the Cabinet in London, though resolved not to allow a new attack on Belgium by Holland, refused to request the King of the Netherlands to reduce his troops. The other French observer, the Duchesse de Dino, supplemented this statement by saying that the protection so far afforded in Belgium rendered it impossible for Grey to continue in office, and that war against France was expected.[63]

A somewhat different comment was supplied from Russian sources. "Les Ministres sont impatiens de nous voir terminer nos affaires", wrote Mme Lieven. "C'est tout sincèrement qu'ils souhaitent que Paskévitch se presse et finisse." Leopold was disliked now as he seemed to side with France. Sebastiani's arguments with regard to the security of Belgium and his reference to "menacing" words of Verstolk, conveyed to London through Granville and Talleyrand, made no difference, and on 15th September, soon after the confidential communication to the British Ambassador, the French Government was made to yield completely: a Protocol, No. 41, "civil" as Palmerston called it, recorded the French declaration; an undoubtedly strange note was struck after six weeks of very irksome negotiations by saying that "the plenipotentiaries requested the Prince of Talleyrand to be persuaded that their Courts would know how to appreciate at its just value the resolution taken by the French Government." [64]

Also on the same day, in Paris, a telegraphic dispatch brought the news of the taking of Warsaw by Russian troops. This was known in London on the 16th; in Paris only after extensive measures of security had been taken, the news was released to the public on the 17th.

V

While the London mission was concentrating on getting everything

[63] Talleyrand to Sebastiani, 4 Sept., *Revue des Deux Mondes* (1910), LVI, 331; same to same, 5, 9 Sept., A.é. Angleterre; Van Zuylen to Verstolk, 6 Sept., R.A.; Greville Diary, 8 Sept., II, 93; Mme de Dino to Thiers, 9 Sept., *Revue de Paris*, IV (1923), 532.

[64] Mme Lieven to Nesselrode, 13 Sept., Lie MSS; Granville to Palmerston, 5 Sept., F.O. France; Sebastiani to Talleyrand, 6 Sept., A.é. Angleterre; Palmerston to Granville, 16 Sept., Bulwer, II, 128-9, where the English translation of the Protocol is also printed.

possible out of the presence of Niemcewicz, the Poles in Paris did not remain idle. As it seemed that little was to be achieved through Sebastiani they tried other channels. From Soult they heard that they could hardly realize how much had been done for Poland. Lafayette promised to approach the King and Périer again. For want of a better solution the envoys wrote even to the Comte de St. Leu – the future Napoleon III, then living in Switzerland – begging him to accept the command of a naval expedition which was about to depart from Le Havre for the Baltic. Fortunately, Louis Napoleon replied that he was fully aware that all governments would equate his name alone with a new wave of revolutions all over Europe, and that, if it were connected with him, the Polish cause would once and for alle be lost in the eyes of diplomats.[65]

Little as it meant to do on its own, the French Cabinet nevertheless was continually applying to Britain to join it in its efforts. Mortemart's report of the serious character of the mutiny in the Russian military colonies was immediately referred to Granville. Renewed instructions were sent to Talleyrand, who spoke about Poland to Palmerston on 4th September, but without success. Palmerston pretended not to share the opinion of French diplomats that Austria might be inclined to favour remonstrances to St. Petersburg concerning Poland. Talleyrand could not understand why the British Ministers, "faisant cependant des voeux pour les Polonais", maintained their inactive attitude. He admitted, however, that the pro-Polish tone of the Press found no reflection in Parliament. Col. Evans' motion was supported by Sir Francis Burdett and O'Connell, but no further action was undertaken. Niemcewicz's next visit to Palmerston was equally fruitless, except for an almost irrelevant assurance by Palmerston, that Prussia would not intervene in case of a Polish victory.[66]

Reports from Poland were, as usual, unreliable, and military reports, for obvious reasons, scarce. The more importance was therefore accorded to other, not always relevant information. Late in August Heytesbury's report of a new levy of troops in Russia

[65] Résumé of an interview with Soult, 26 Aug., Lafayette to the Paris envoys, 29 Aug., Platter to the Comte de St. Leu, 28 Aug., Louis Napoleon Bonaparte to Plater, 8 Sept., B.P.

[66] Granville to Palmerston, 29 Aug., 3 Sept., F.O. France; Mortemart to Sebastiani, 17, 20 Aug., A.é. Russie, cf. above, p. 233. Talleyrand to Sebastiani, 4 Sept., A.é. Angleterre; *Hansard*, VI (3rd Series), p. 1216; Niemcewicz to the Paris envoys, 13 Sept., B.P.

arrived in London. Chad wrote that he knew on good authority, that the Russians had no suitable artillery to attack Warsaw. Flahaut reported that the King of Prussia was strongly impressed by an address from the City of Köningsberg to the effect that intercourse with the Russians was censured as being the source of spreading of the cholera epidemic. In its relations with foreign missions, however, the tone of Berlin changed a good deal. The question of passports to Poland had become irrelevant, said Ancillon, since Warsaw was cut off from the West by Russian forces. The King of the Netherlands would not be allowed to reopen hostilities; yet, pressed for the recognition of Leopold which had been promised by the Conference in July, the Prussian Minister said that Austria and Russia should take the lead in this matter. When Chad repeated Palmerston's protest with regard to Prussian supplies for Paskevitch's army, Ancillon retorted that Prussia only allowed the private export of provisions, whereas arms and munitions were arriving from England, without which the Polish revolt would have been suppressed long before. And yet, on the same day Chad reported the unexpected news, which even got into the "Preussische Staatszeitung", that the Russians were retreating from before Warsaw. Only a few days later this appeared to have been partly true, and on the 11th Chad sent his first account of the taking of Warsaw.[67]

This information was not what was expected and the circumstances were entirely different from those at the time of the battle of 25th February. None of the interested parties had foreseen such a speedy outcome, and that explains perhaps why at that time the Poles were still able to secure quasi-diplomatic intervention on the part of Austria, an intervention counselled by many, and which had been long diligently sought.

During July and August the Polish agent Count Andrew Zamoyski was once more in Vienna. It was still thought that Metternich and Gentz were favourably disposed towards Poland, and the agent again managed to be introduced to the Chancellor in great secrecy. In an interview on 19th August Metternich assured him that it was too late for any intervention, even for an

[67] Heytesbury to Palmerston, 18 Aug., F.O. Russia; Chad to Palmerston, 28, 31 Aug., 3, 4, 9, 11 Sept., F.O. Prussia; Flahaut to Sebastiani, 21 Aug., A.é. Prusse. When he sent his request for recognition of Leopold to Prussia, Palmerston did not expect success (to Granville, 26 Aug., Bulwer, II, 121).

offer of submission to be made through the good offices of the Emperor Francis; some days later, however, he agreed to give Zamoyski a letter containing the Emperor's advice to submit. The Russian Ambassador concurred in this step, which in itself was evidence of the eagerness of his Court to come to terms, and a secretary of the Russian Embassy accompanied Zamoyski to Warsaw, where, unfortunately, they arrived after the entry of the Russians. Answering the British *Chargé d'affaires'* question concerning this journey Metternich said that Russia "was as convinced as himself of the importance of not proceeding to extremities." From other sources it would appear, however, that Metternich, having always tried in vain to step into the Polish question, seized this opportunity as his last chance.[68]

It may well be assumed that, if Zamoyski's overture stood a chance of success at Warsaw, some negotiations with Paskevitch might at least afford a breathing-space to the Poles. For at St. Petersburg as late as 14th September, that is after the actual entry of Paskevitch into Warsaw which took place on the 8th, it was feared that the arrival of Ramorino's corps in Warsaw might render the capture of the city impossible. Paskevitch possessed instructions which ordered him, after one unsuccessful attack, to retreat westwards to winter quarters supported by supplies from Prussia.[69]

Lastly, whatever the official attitude, pro-Polish feeling existed in Prussia not only among the liberal bourgeoisie but in higher circles as well, and this, as has been shown, was not without influence upon the Government.[70]

[68] A. Zamoyski, *My adventures,* II, pp. 46-7, 58, 79-86, also p. 102: Gentz told Zamoyski on 20 August: "Depuis plusieurs mois déjà, le Prince Metternich a proposé une entrevue entre les trois souverains, vos voisins, pour l'affaire de la Pologne, après que la crise actuelle aura été étouffée; l'Empereur Nicolas s'y réfuse."; Forbes to Palmerston, 6 Sept., F.O. Austria. The French Ambassador in Vienna believed that war was just as possible because of Belgium as because of Poland, and that Metternich had had it in mind when he had refused to support the request to the King of the Netherlands to reduce his army. Maison to Sebastiani, 5 Sept., A.é. Autriche.

[69] Heytesbury to Palmerston, 14 Sept., F.O. Russia; Tokarz, *The Polish-Russian War in 1831,* p. 609.

[70] F. v. Gagern to his son, 9 Sept.: "In Berlin weiss niemand wer Koch oder Keller ist. Der König ist Russisch und die Nation Polnisch." A. v. Gagern, *Das Leben des Gen. Friedrich von Gagern,* 1856-7, II, 128; Baroness von Bülow, the wife of the Prussian Minister in London wrote to her sister on 31 August on Matuszewic; "a Pole, but not of those Poles for whom my heart beats so warmly", and again, "un-Polish Pole like this Matuzewitz"; on 15 September, commenting

Under the circumstances, the first accounts after the taking of Warsaw represented the situation as if it had not been a definite blow to the Poles. Chad suggested that the fall of the city was due to the concealed treason of the Polish commander. The British *Chargé d'affaires* in Vienna reported that the Polish army which was retreating northwards was still strong and capable of action.[71]

When the fall of Warsaw was known in London Talleyrand observed "une douleur profonde; cependant personne ne désespère ici; Varsovie n'est pas la Pologne." This opinion was not shared by Palmerston: the Polish case "had become for some time hopeless", he wrote to Granville. "If they could not defend Warsaw, what can they do at Modlin or Plock?" [72] Greville noted that since the deposition of Skrzynecki – known in London for some weeks – sympathy for Poland had decreased, while Mme Lieven delivered a comprehensive comment: "Les Ministres sont fâchés et honteux. Lord Grey qui [sic] se montre convenablement, mais je crains que cela ne soit par le sentiment personnel plutôt que politique. Le public est assez froid, les Torys sont dans l'enchantement et la presse est détestable. Voilà où nous en sommes au second jour de la nouvelle de la prise de Varsovie."[73]

Another important commentator, Wellington, said that the Polish revolt had shown the weakness of Russia, "which was incapable of any great independent effort." But this would make the French more tractable; on the other hand, Prussia and Austria had been financially so exhausted by military expenses "that they could hardly hold out a month longer."[74]

The consequences of the breakdown of the Polish movement were almost immediately felt in the various capitals. Flahaut reported

---

upon the taking of Warsaw: "I quite see that it is to Prussia's advantage to settle this Polish business, but my heart bleeds for Poland and I grudge Russia the triumph!" Gabrielle von Bülow, *Memoir compiled from Family Papers*, 1897, pp. 204-6.

[71] Chad to Palmerston, 13 Sept., Forbes to Palmerston, 15 Sept., F.O. Prussia and Austria.

[72] Talleyrand to Sebastiani, 17 Sept., A.é. Angleterre; Palmerston to Granville, 16 Sept., Bulwer, II, 127. The Polish Army retreated towards Plock but it was not clear whether in consequence of Paskevitch's instructions or to prepare for further action.

[73] *Greville Diary*, 17 Sept., II, 197; Mme Lieven to Nesselrode, 20 Sept., Lie MSS.

[74] A. Aspinall, *Three Early XIX Century Diaries* (Ellenborough Diary), 1952, p. 128.

renewed remonstrances by the Prussian Government for the withdrawal of the French forces from Belgium (they did not yet know of any arrangement) and for the maintenance of Maastricht and Venlo within the new frontiers of Holland. At The Hague both La Rochefoucauld and Bagot feared that the difficulties, raised by William I, would be increased by the renewed hopes he might entertain with regard to the improvement of his position in the international sphere after the victory of the Russians. Nor did Palmerston fail to observe that the Belgians would regret their opposition to the negotiations, because delay thus caused put them in an unfavourable position so that they would begin at an unpropitious moment "since the success of the Russians gives advantage to the King of the Netherlands." [75]

The Dutch plenipotentiaries realized this immediately and Van Zuylen seized the opportunity of an incidental call on Lieven, just after the latter had received the happy news, to ask for a new effort by Russian diplomacy in their favour. He noted, somewhat sourly, "plenty of delightful promises". But even less was achieved with Bülow, and the plenipotentiaries felt slightly uneasy about any new prospects. A few days after the receipt of the Warsaw news they devoted a major part of their dispatch to the influence of the Polish question upon the work of the Conference. They began by stating that the Press and society were surprised at the news of the surrender of Warsaw, as were the Ministers, especially Palmerston and Grey, the more so because until recently they had been convinced that Poland was lost to Russia. "Bien que personne ne se dissimule que cet événement doive nécessairement influer sur la négociation qui nous concerne", the dispatch went on, "il est évident que l'on veut ne pas se l'avouer." Yet, after receiving the Warsaw news the Russian plenipotentiaries declared to Palmerston and Talleyrand that this event would not change Russia's policy in the Belgian question. This declaration, the Dutchmen believed, constituted an acknowledgement to Britain in return for her cautious attitude in the Polish question, which had made France abstain from a more active policy in this matter. On the other hand, as the representatives of Russia, Austria and Prussia constantly maintained that for them war in Poland

[75] Flahaut to Sebastiani, 16 Sept., Guizot Papers; Bagot to Adair, 17 Sept., Bagot MSS; La Rochefoucauld to Sebastiani, 16 Sept., Ged., II, 258-9; Palmerston to Granville, 16 Sept., Bulwer, II, 129.

was the main obstacle to their adopting a positively pro-Dutch line at the Conference, Falck and Van Zuylen seized the present opportunity to remind them that, the main obstacle having been removed, the just claims of the King of the Netherlands should now be supported.[76]

After his verbal declaration Lieven called on Talleyrand again, to renew his former declaration, made after his return from Russia just after the news of the outbreak of the revolution in Poland, with regard to his Court's desire to maintain friendly relations with France. This gesture impressed Talleyrand who, considering the radical change brought about by events in Poland, attached great importance to the words of the Russian Ambassador: "elles... peuvent contribuer à rassurer sur les intentions que quelques personnes paraissent encore attribuer à la Russie." [77]

At the same time Bagot was afraid that there were political motives behind the congratulations which the Prince of Orange received from the Emperor of Russia; early in September Esterhazy received Metternich's letter with the information that Nicholas did not conceal that he would have helped the King of the Netherlands in his campaign had he not been handicapped by the Polish revolt. This might not have been widely known, but the British Ambassador preferred to comment on the Prince's departure for the army, with no intention to recommence hostilities: "as he said." No wonder that the utterances of the "Journal de la Haye", considered as a semi-official organ of the Netherlands Government, increased the apprehensions of the French. Sebastiani referred to the increased warlikeness of this newspaper and stressed to Talleyrand the importance of presenting a firm opposition to any attempts at concessions in favour of Holland.[78]

The person most of all affected was King Leopold. He could only partially rely on France, as at that stage the prolonged occupation of Belgium by French forces was causing a conflict between France and England, his most obvious ally. Giving up French protection therefore, he wrote to Grey in what seemed to be the utmost despair: "I beseech you,... do what is right by Holland

[76] Falck and Van Zuylen to Verstolk, 17, 20 Sept., R.A. See Appendix XIII; Stockmar to Leopold, 26 Sept., *Memoirs of the Baron Christian Frederick von Stockmar*, ed. E.v. Stockmar, 1913, I, 196.

[77] Talleyrand to Sebastiani, 24 Sept., A.é. Angleterre.

[78] Bagot to Palmerston, 23 Sept., Bagot MSS; Arneth, *Johann Freiherr von Wessenberg*, II, 123-4; Sebastiani to Talleyrand, 21 Sept., A.é. Angleterre.

but do not maltreat this country which really does not deserve it now, whatever it might have done before." This letter, written before the news of the fall of Warsaw, ended with an expression of Leopold's belief that the Russians "will do everything in the world to keep this negotiation open till they can come into play." [79]

Like the Belgian King, Louis Philippe also foresaw great difficulties menacing the new State. As the Conference abstained from requesting the King of the Netherlands to reduce his forces, while no peaceful assurances could be obtained at The Hague, it did not seem unlikely that the Belgians might feel provoked to new hostilities, which from Holland would derive even more advantages than from the August Campaign; reports from The Hague of the way in which the surrender of Warsaw was considered, could not fail to alarm the French mission. The "Journal de la Haye" wrote: "Cet événement était immense par le résultat qu'il doit avoir sur les affaires de l'Occident de l'Europe, par la liberté d'action qu'il rendait à la Conférence de Londres." Accordingly, William I was believed to contemplate a new invasion of Belgium and an Orangist revolution there. Besides, whoever might be better informed about the state of affairs in London, The Hague could rely on the "Journal", for Gourieff, the Russian Minister, affirmed that an application for assistance to Russia would not be made in vain, "et que, rendue à toute liberté d'action par la soumission de la Pologne, la Russie ne peut manquer d'appuyer de sa puissante influence la cause du Roi Guillaume dans la Conférence de Londres." [80]

There is ample evidence that these words had not been spoken in vain. While the Conference resumed its work on the new Treaty proposals, an extension of the armistice which was to expire on 10th October was badly needed. Palmerston agreed with Grey to send "a few ships" to the Scheldt to impress the Dutch, but at the same time noted that "Bülow, Wessenberg and the Russians make themselves partisans" of the Dutch cause. In the meeting on 28th September Bülow "vehemently" opposed an extension of the armistice; Palmerston was unable to make the Eastern representatives

[79] Leopold to Grey, 10 Sept., Howick MSS. This letter is quoted above, p. 222.

[80] Louis Philippe to Sebastiani, 22 Sept., *Revue des Deux Mondes,* LVI (1910), p. 325; Adair to Palmerston, 23, 26 Sept., Howick MSS; La Rochefoucauld to Sebastiani, 28 Sept., Ged., II, 260.

agree to an extension of the armistice for one month; all that could be achieved was to request Van Zuylen to ask for a date at The Hague, but no more than two weeks' extension was thereby anticipated.[81]

On the same day Palmerston informed Grey that Talleyrand requested a special Cabinet meeting on Polish affairs, to consider what steps should be taken in defence of the Polish institutions. Though it had been clear that the French Government experienced the greatest internal difficulties on account of Poland, Grey had some criticism of its Polish policy. Some days before Greville, informing him of the depressed feeling at the French Embassy, was told that "the French Ministry had been very imprudent about Poland." Greville was surprised: there was no access to Poland except through Prussia. Grey then continued: "They might have sent a fleet to the Baltic with our concurrence though we could not urge them to do so." In view of the Paris events Grey expected that a "ministry of Lafayette, Lamarque and all that party" would lead France into war, whereas "Russia.... powerless, crippled by this contest and under the necessity of maintaining a great army in Poland", and Austria, and Prussia, also weakened and with a population liable to be revolutionized by the French, accounted for a fair chance of success for France in such a contest.[82]

Things looked black in Paris those days, but Grey's worst expectations were not realized. There were riots at the news of the taking of Warsaw. Granville reported on the 16th that they were under control, but on 19th September he regretted to say that "the state of things is very alarming, a general impression seems to prevail that some change is impending. The Ministers seem to have lost all influence in the Chamber of Deputies." The National Guard, unlike on other occasions, dealt very reluctantly with the rioters. The Russian Ambassador was requested by the police to leave his house, because they were unable to protect him. "On s'attend à des séances orageuses et difficiles pour les ministres", wrote the Dutch Envoy, as Lafayette requested the Government to furnish an explanation with regard to a written assurance given to

[81] Palmerston to Grey, 26, 28, 29 Sept., Howick MSS.

[82] Greville Diary, 20 Sept., II, 200; Niemcewicz wrote to Paris that when he had asked Grey why he had not sent a fleet to the Baltic, Grey blushed. Niemcewicz to the Paris envoys, 20 Sept., B.P.; cf. also J. Stuart Reid, *Life and Letters of the first Earl of Durham,* 1906, I, 308.

the Polish mission that Poland would obtain assistance within two months.[83]

On 21st September Granville reported the restoration of law and order. This relieved the anxiety of Grey: if Périer's Cabinet remained in office, the peace of Europe was saved.[84]

VI

There was considerable delay in the progress of the work of the Conference owing to the unco-operativeness of the Belgians, and the protest against this, officially made at Brussels by Adair and Belliard, coincided with the Warsaw news and the first reactions to it. In the new negotiations, therefore, the tide turned positively in favour of Holland. Talleyrand, who was now taking a much lesser part in the work than he had done before, did not particularly like the tone adopted by the Russian plenipotentiaries, but was satisfied that their hostility would not exceed "quelques formes d'arrogance." Reports from Berlin and Vienna confirmed that the aftermath of the Polish war was still felt; Prussia was full of refugees, and so was the Republic of Cracow, the temporary occupation of which by Russian troops, greatly disapproved of by Bernstorff, was the subject of talks held in Berlin in great secrecy; the Dutch Minister there was confused into believing that it concerned "une déclaration énergique relativement à la Belgique et au Luxembourg." [85]

On 15th October the Conference produced a new project for arranging Belgian affairs, hereafter known as the 24 Articles. They were largely based on the XVIII Articles of June, but the ambiguous points were fully worked out and put in favour of Holland.

[83] Granville to Palmerston, 16, 19 Sept., F.O. France; *Journal du comte Rodolphe Apponyi,* 1913–1926, ed. E. Daudet, II, 58; Fagel to Verstolk, 19 Sept., R.A. Reference is made to the interview of 7 July; the dispatch to Warsaw was written by the Poles, but the courier obtained a French passport and was paid by Sebastiani. Cf. above, p. 163 and 179.

[84] Granville to Palmerston, 21 Sept., F.O. France; Grey to Mme Lieven, 23 Sept., *The Lieven-Grey Correspondence*, II, 281.

[85] Adair to Palmerston, 19 Sept., Palm MSS; Talleyrand to Sebastiani, 12 Oct., A.é. Angleterre, cf. Webster, I, 142; Forbes to Palmerston, 11 Oct., Chad to Palmerston, 12 Oct., F.O. Austria and Prussia; Perponcher to Verstolk 20 Sept., R.A. The mission of General Clam- Martinitz to Berlin was the subject of speculations in various dispatches at the end of September and throughout October until it appeared that it concerned Cracow; the appointment of General Krauseneck to confer with Clam must have meant a great deal to Perponcher.

Both Bagot and Palmerston, therefore, the latter on the authority of Van Zuylen, at first believed that Holland would now agree. In the secrecy of the Council Verstolk, the Minister for Foreign Affairs, advocated acceptance, arguing, as another Minister, Van Maanen, had done a fortnight before, that the three Powers friendly to Holland continued to profess their intention to follow England.[86]

At the same time Palmerston did not overlook measures of security. As the armistice by then had officially expired he instructed Bagot to make it quite plain in The Hague that the renewal of hostilities would not be tolerated but news poured into London, indicating not only the Dutch preparations, but also movements of Prussian troops on the Belgian frontier, while Bernstorff refused to consider a reduction in the Prussian armed forces as long as Belgian affairs remained unsettled. Eventually, the Foreign Secretary thought it advisable to inform Van Zuylen that the squadron of Admiral Warren, now put in a state of readiness, could act on direct orders from the British Ambassador in The Hague.[87]

The Belgians acquiesced in their defeat and subscribed to the new project; King Leopold considered himself exploited by the Powers, and to Adair as well as to others spoke of abdication. Granville regretted that the Minister for Foreign Affairs, introducing the project in the Belgian Chamber, explained the modification in favour of Holland as the result of the failure of the revolutionary movements in Poland and Italy. The French Chambers, he thought, would gladly seize the argument to make trouble.[88]

At The Hague the situation soon "cleared" and the British Ambassador observed the usual hesitant and speculative attitude. On 7th November he was positive about William I's refusal, in which he would be supported by his subjects. "The King believes", Bagot wrote, "that he has the means of maintaining the burden

[86] Bagot to Palmerston, 18, 22 Oct., Palmerston to Bagot, 15, 18 Oct., Bagot MSS; Memorandum by Verstolk for the Council of Cabinet, 22 Oct., Memorandum by Van Maanen,2 Oct., Van Zuylen and Van Maanen Papers respectively, R.A.

[87] Palmerston to Bagot, 15 Oct., Bagot MSS; Holland to Grey, 30 Oct., Howick MSS; Adair to Palmerston, 21 Oct., Palm MSS; Chad to Palmerston, 16 Oct., F.O. Prussia; Palmerston to Van Zuylen, 1 Nov., Palm MSS.

[88] Adair to Palmerston, 19 Oct., Palm MSS; Granville to Palmerston, 24 Oct., F.O. France.

of his present position for two years longer, and perhaps for more."[89]

Meanwhile the project sent to the Courts friendly to Holland was being considered by them. Ancillon wrote to the Prussian Minister at The Hague to warn the King that in case of his refusal coercive measures might be applied by England, France and Prussia jointly. He told Chad that William I had nothing to expect, but that the object of his delay was to see what would be the reaction of his friends, and this delay was liable to be prolonged extensively in view of the distance between St. Petersburg and the capital of Holland. As the interview went on, the Minister revealed "the secret thought" of the Prussian Government: the Republicans, the Orangists, and the French Party, he said, together with his unsettled international position, would present Leopold with so many difficulties that he might abdicate, thus giving an opening for a partition of Belgium.[90]

There remained Vienna and St. Petersburg. Metternich was at first not opposed to the new arrangement. There was, as usual, the greatest delay in news from Russia. Heytesbury delivered a peculiar introduction to what was to be expected from the Imperial Cabinet. He heard of Nesselrode's approval of Lieven's declaration to Palmerston and Talleyrand that Russian policy would undergo no change "in consequence of the success which had attained the Emperor's arms" in Warsaw; special reference was made to the moment chosen by Lieven. The British Ambassador heartily concurred: "The policy of this Government.... is at this moment in the strictest sense *conservative,* not aggressive." Nothing could play better into William I's hands; meanwhile, at the end of October Nicholas left for Moscow, partly at least with the intention of avoiding diplomatic intervention in favour of Poland, and communication with the Russian Government became even more difficult.[91]

On 15th November, at Bagot's suggestion, the representatives of the Five Powers in The Hague called on the Minister for Foreign Affairs to demand from him an official reply to the Conference's proposals. The British Ambassador reported: "In the course of our conversation the Ministers of Austria and Prussia and the *Chargé*

[89] Bagot to Adair, 22 Oct., to Palmerston, 23, 25 Oct., 1, 7 Nov., Bagot MSS.

[90] Chad to Palmerston, 7 Nov., F.O. Prussia.

[91] Forbes to Palmerston, 1 Nov., Heytesbury to Palmerston, 12 Oct., F.O. Austria and Russia; Bourgoing to Sebastiani, 9 Nov., A.é. Russie.

d'affaires of Russia severally took occasion to express in the strongest manner to M. de Verstolk the entire and unqualified adhesion of their respective Cabinets to every part of the arrangement now determined by the Conference." [92]

The Conference, however, did not await the effect of this and similar steps undertaken by members of the diplomatic corps at The Hague, and on 15th November the delegates of the Five Powers signed the Treaty of 24 Articles with the King of the Belgians. In December the King of the Netherlands sent his official refusal to agree to the October project of the Conference.

The ultimate date for the ratification of the Treaty which commanded "the entire and unqualified adhesion" of its signatories was fixed for 15th January, 1832; but it was not before May of that year that the three Eastern Powers sent their consent supplemented with reservations which angered the other two, and troubled King Leopold. Part of the proceedings which preceded these steps will be examined here.

As soon as the Treaty had been signed, Palmerston stressed to Chad the importance of Prussian ratification. "The nearness of Berlin to Brussels enables the Prussian Government to take lead in this measure and I am convinced that such a measure would have a powerful effect upon the determination of the King of the Netherlands." [93]

The news of the Treaty having been signed arrived in Berlin on 24th November. Ancillon regretted that no time had been allowed for Nicholas to dispel "les dernières et chimériques espérances" of William I. Nicholas, who had declared long ago that he would consent to no agreement which had not been previously accepted by William I, might feel offended at an attempt to force his hand and might refuse to ratify it. The King of the Netherlands had not yet been able to learn Nicholas's opinion about the project itself. "Il est à croire que M. de Nesselrode serait sacrifié," as Bresson, now *Chargé d'affaires* at Berlin, hastened to sum up a view spreading in the Prussian capital.[94]

To Chad Ancillon said that Bülow's powers were not extensive enough to allow him to sign such a treaty, and if Prussia chose to

[92] Bagot to Palmerston, 15 Nov., F.O. Holland.
[93] Palmerston to Chad, 23 Nov., Palm MSS.
[94] Bresson to Sebastiani, 27 Nov., 2 Dec., A.é. Prusse.

take the formal view, she might refuse ratification on that ground alone.[95]

No apprehensions with regard to the career of Nesselrode were entertained at the Prussian Embassy in London. Early in December Bülow told the Dutch plenipotentiaries that Prussia advocated acceptance of the Treaty; and as Russia needed Prussian assistance for her policy in Poland, she would not alienate Prussia by adopting a different line in the Low Countries. The Minister also told Talleyrand that the Prussian ratification was on its way to London. The French Ambassador believed the same about the Russian and Austrian documents, and tried to calm Paris down, whence the application of coercive measures against Holland had been put forward. After Bülow had shown him his dispatches, he became quite positive about the decision of the Prussian Government and more assured of that of Russia. Prussia inclined to follow Russian policy in Poland, where Russia was supposed to have predominant interests, the document said, whereas Belgium was far more a concern of the Berlin Government than one of St. Petersburg, and the latter would certainly appreciate that fact.[96]

Yet, on the receipt of the refusal of the King of the Netherlands, which arrived in London on 14th December, even Grey thought that Russia had encouraged it. No wonder that Palmerston had to reckon with the French Government as well as King Leopold fearing an attempt at a new invasion of Belgium or at an Orangist revolution there.[97]

Then the news from Russia began to arrive. Before he had seen either the project or the Treaty, Nesselrode warned the Ambassador in London against consenting to "des mesures trop fortes" against William I. The first news obtained from Berlin read: "L'Empereur ne consentira jamais à des mesures coercitives contre le Roi de Hollande"; but Nicholas advised King William to accept the project of the treaty – the 24 Articles themselves – as a basis for negotiations. Bernstorff then commented to Bresson that this meant that Nicholas had renewed his former declaration – as had been apprehended by Ancillon – and that the Treaty itself was in danger:

[95] Chad to Palmerston, 2 Dec., F.O. Prussia.

[96] Van Zuylen to Verstolk, 8 Dec., R.A.; Sebastiani to Talleyrand, 28 Nov., Talleyrand to Sebastiani, 30 Nov., 5, 14 Dec., A.é. Angleterre.

[97] Grey to Mme Lieven, 14 Dec., *The Lieven-Grey Correspondence,* II, 304; Palmerston to Bagot, 16 Dec., Bagot MSS.

non-ratification by even one of the parties concerned would invalidate it.[98]

The British Minister also heard a statement which indicated some hesitation. On hearing the early Russian news Ancillon told Chad that Prussia would ratify, but not before the others had done so, and in the meantime would employ her persuasion to make them agree.[98]

Meanwhile, instructions from Russia sent after the news of the October project, could only strengthen the impression of Russian partiality for the King of the Netherlands. Nicholas not only deplored the fact that his representatives had agreed to the 24 Articles but instructed them to defend the King against coercive measures which might be applied against him if he refused this new project and reopened hostilities: the Conference should in such a case proclaim its neutrality! "La cause du roi Guillaume est trop juste pour être laissé sans défense; sa dignité trop élevée pour être méconnu, enfin son infortune trop grande pour ne commander le respect," wrote the Vice-Chancellor to London.[99]

Of these instructions little seems to have penetrated into diplomatical circles in London except the non-ratification, of which there could be no doubt. Heytesbury's report of his interview with Nesselrode after the latter had learned about the Treaty breathed moderation. The Vice-Chancellor merely said that the Conference acted too hastily, and he first wanted to know William I's opinions. And indeed, the official answer of Nicholas to the King did not avow any other object in his delay in ratification than to show deference to the wishes William I might have with regard to the Treaty. "...je ne saurais nourrir l'espoir de faire adopter la même résolution par mes alliés", he wrote. "Dans cet isolement, je n'aurai offert à V.M. qu'une preuve nouvelle de l'amitié que je lui porte, sans pouvoir me dissimuler qu'elle aura été en quelque sorte stérile si elle ne servait pas à lever les obstacles qui s'opposent encore à un arrangement à l'amiable." The mission of Orlov, Nicholas' principal aide-de-camp, to whom the King would have to make clear his objections to certain points of the Treaty, had for its main object to explain to William I the uselessness of further delay.[100]

[98] Nesselrode to Lieven, 12 Nov., Martens, *Recueil,* XI, 461; Bresson to Sebastiani, 8, 11 Dec., A.é. Prusse; Chad to Palmerston, 11 Dec., F.O. Prussia.
[99] Nesselrode to Lieven, 29 Nov., Martens, *Recueil,* XI, 462.
[100] Heytesbury to Palmerston, 10 Dec., F.O. Russia; Nicholas to William I,

Months had to elapse, before this determination of Nicholas was fully understood by the other parties of the Conference. In the privacy of his study Nicholas went even further; in a personal memorandum for his own use on the international situation, he noted his intention to postpone recognition of Leopold – though he was far from agreeing with William I's political methods – but he admitted the impossibility for Russia to oppose any coercive measures with regard to Holland, should the Western Powers resolve to apply them.[101] In December, however, nothing but the refusal of Russia was known, and this caused great confusion in Berlin and London.

On 17th December the British Minister called on Ancillon to talk about Portuguese affairs but found him extremely embarrassed by the Russian refusal to ratify the Treaty. Bernstorff at first believed that this refusal constitued another mark of deference towards the King of the Netherlands, as Nicholas only learned of the conclusion of the Treaty after he had advised William I to subscribe to the October project and before he knew his opinions upon it. But Ancillon surprised Chad by stating that not only did he consider the refusal absolute but that in view of the dissent of one of the contracting parties the Treaty itself became null and void. Some days later Bernstorff also arrived at the same conclusion.[102]

In London the negotations between Belgium and the Four Powers with regard to the fortresses had reached a critical point and much harm was expected by the delay in ratification. Both Russian plenipotentiaries were rather worried. Nicholas's advice to The

28 Dec. 1831, 30 Jan. 1832, Ged., III, 516, 525; Bourgoing to Sebastiani, 28, 31 Dec., A.é. Russie.

[101] Martens, *Recueil,* VIII, 167-170. The Editor erroneously puts the date of the memorandum as "last days of 1830", for which 1831 must be read. This document reflects the confusion in the mind of Nicholas about the Belgian affairs: "..Nous avons reconnu le fait de l'indépendance de la Belgique, parce que le Roi des Pays Bas l'a reconnu lui-même (?), mais ne reconnaissons pas Léopold, car nous n'avons aucun droit de pouvoir le faire. Mais en même temps ne cachons pas notre désapprobation manifeste de la conduite double et fausse du Roi, et retirons-nous de la conférence. — Si la France et l'Angleterre s'unissent pour tomber sur la Hollande, nous protesterons *car nous ne pouvons faire plus* [italics mine]; mais du moins le nom Russe n'aura pas été avili par la complicité d'un pareil acte." The Editor himself was puzzled by the reference to the recognition of the Belgian independence by William I, which could not be true at the end of 1830, though at the end of 1831 might be based on the "Bases de séparation".

[102] Chad to Palmerston, 18, 26 Dec., F.O. Prussia; Bresson to Sebastiani, 17, 20 Dec., A.é. Prusse.

Hague with regard to the 24 Articles was known; Talleyrand did not believe in an absolute refusal of ratification by Russia. He advised his government, therefore, not to press the matter of the fortresses which necessarily divided Britain and France: the Polish question united the three Eastern Governments and it was then of great importance for France to continue on the best possible terms with Britain. The ratifications were anxiously awaited in London.[103]

This situation affected the French Cabinet most of all, while the further reports made neither the Russian intentions nor those of the other two Eastern Powers clear. Bresson, with his usual haste, concluded that the advice given to William I meant nothing when accompanied by delay in ratification. He attached importance to the opinion of the Grand Duchess Helen, at that time visiting Berlin after a short stay at The Hague, that Lieven and Nesselrode might fall in disgrace. Sebastiani professed to be offended because Nicholas had explained his delay to all members of the Conference except France. Maison reported the refusal of Vienna, though Metternich would not admit to him that he did not dare to ratify alone. The Chancellor sent a letter of strong disapproval to London on account of the part that Wessenberg and Esterhazy had played in accepting the Treaty. Ancillon admitted the likelihood of new complications about Belgium; he regretted, however, that Prussia could do nothing without reference to St. Petersburg. However, Chad believed that the prospects of war and similar consequences, which had been under serious consideration in Prussia as possible results of the failure of the Belgian Treaty, might prove successful arguments for ratification with some members of the Prussian Government and certainly with the Minister for Foreign Affairs, who did not conceal his embarrassment. "Je tremble de recevoir des nouvelles de Londres", he told Chad, "je crains des mesures violentes." [104]

However, no energetic measures could have been contemplated in London. The Convention on the Belgian fortresses had just been signed and could not but contribute to the irritation of public opinion in France; it was believed in London that in the question of

[103] Talleyrand to Sebastiani, 20, 27 Dec., A.é. Angleterre, to Mme Adelaïde, 27 Dec., *Nouvelle Revue Rétrospective*, (1902) XVI, 57.

[104] Bresson to Sebastiani, 23, 24 Dec., A.é. Prusse; Sebastiani to Bourgoing, 30 Dec., A.é. Russie; Chad to Palmerston, 25 Dec., F.O. Prussia, 28 Dec., Palm MSS; Maison to Sebastiani, 27 Dec., A.é. Autriche; Metternich to Wessenberg and Esterhazy, 26 Dec., Arneth, *Johann Freiherr von Wessenberg*, II, 134.

the ratification of the Treaty of 15th November the existence of Périer's Government was again at stake. A mere reference to any coercive measures against William I could but too easily excite the French, who would be happy to use any pretext to enter Belgium. Much as the British Government was affected by the delay in the ratification it could not think of trying to speed up matters by force.[105]

Hence Palmerston's violent letter to Chad, full of anger at Ancillon, having "upon the first rumour of the refusal of the Emperor Nicholas" eaten all his words concerning Belgium. "I see nothing that can prevent a war except the ratification of the Treaty", wrote the Foreign Secretary. "If Prussia refuses, one or other of the two things will be demonstrated: either that in spite of her pacific professions Prussia is panting for war, or else that she is not an independent agent and is dragged by Russia at her chariot wheels. You must insist upon having an immediate answer, yes or no, will Prussia ratify or not, and it should be given in a Prussian note in order that it may be produced, if it should become necessary to do so, if the Papers are laid before Parliament, which upon all occasions of unsuccessful negotiations [they] invariably are." [106]

It was clear that Palmerston lacked an argument; his reasoning did not question any of the points raised in Berlin, and besides, if Prussia – or France – would really not shrink from war, it was equally clear that Palmerston's diplomacy in forcing the Treaty upon the Conference only hastened the moment from which there was no withdrawal. On the other hand, the state of France might well be seen with different eyes from Berlin, and this, according to Chad, was very much the case. Threats held out from London would lack force.[107]

Palmerston's concluding sentences ran as follows: "Tell M. Ancillon that we are beginning to regret that we did not take a more decided part in Polish affairs before the capture of Warsaw, we might have done some good in Poland and have less trouble with this Treaty."

[105] Talleyrand to Périer, 2 Jan., 1832, Périer to Talleyrand, 9 Jan., Talleyrand, *Memoirs*, IV, 383, 386; Grey to Mme Lieven, 27 Jan., *The Lieven-Grey Correspondence*, II, 327; Mme Lieven to Nesselrode, 14 Jan, 1832, Lie MSS.

[106] Palmerston to Chad, 30 Dec., Palm MSS.

[107] Chad to Palmerston, 28 Jan., Palm MSS.

At that time those words hardly had the air of a menace.

The last months of 1831 and the beginning of 1832 witnessed some diplomatic activity with regard to Poland, directed towards Russia as well as towards Prussia, where refugees presented a considerable problem; if on no other account, financial reasons were enough to make the Prussian Government deliver the people in question to the Russian authorities. Otherwise the most prominent of the refugees made their way to England and France. Among them was Prince Adam Czartoryski, who narrowly escaped arrest when the Russian forces unexpectedly invaded the Republic of Cracow. On the express orders of Metternich he was supplied with an Austrian passport and escorted into Austrian territory.

Czartoryski's visit to London in the beginning of 1832 created a slight diplomatic incident between the Russian Embassy and the British Government; Grey in a more intimate and undiplomatic way commented upon the effect of the Polish defeat. "It is really heart-breaking to see him", Grey wrote to Brougham in his New Year letter, "& now these d....d Russians are doing all they can to throw the whole Belgian affair into confusion. It is to be regretted that we had no power of sending a fleet into the Baltic last summer to settle the matter of Poland."[108]

[108] Cf. Webster, I, 190; Grey to Brougham, 1 Jan., 1832, *Lord Henry Brougham, Life and Times, written by himself*, 1871, III, 164-5.

CONCLUSION

The year 1832 began with the question of the ratification of the Belgian Treaty still unsettled; the King of the Netherlands continued refusing even to consider the recognition of Leopold's kingship in Brussels. William I was, however, unable to make the special envoy of Nicholas, Count Orlov, a partisan of his cause. Orlov spent some weeks in The Hague during February and March, 1832, trying to convince the King of the necessity of coming to terms with Belgium. He had no success, but neither did the King succeed in winning the Russian envoy to endorse his own standpoint: that he had never given up the sovereignty of Belgium. Nevertheless, the King's failure with Orlov did not prevent the latter's Imperial master to send in his ratification of the Treaty with the reservation that "the consent of the King of Holland must be obtained to the question on which he was still in dispute with the Conference" This was a very meagre result for William I, but in the meantime Palmerston himself had had time to become convinced that the Belgian Treaty needed revision – after he had succeeded, with considerable difficulty, in getting the Belgian delegate to agree to accept, in April 1832, the Russian ratification so reserved.[1]

King Leopold, a more experienced man, saw a better chance of settling the affairs of his new kingdom. In the beginning of April news came to the West of Europe of the announcement in Warsaw of the "Organic Statute" for the Kingdom of Poland in lieu of its former constitution; Leopold equated this measure with the incorporation of the Kingdom into Russia, which, in his opinion, would please neither Paris nor Berlin or Vienna. "It is a great event from which a great deal may be made in favour of the conclusion of the Belgian business", he wrote to the Belgian Minister in Paris. "This act of Russia is contrary to all stipulations;

[1] Smit, p. 174-5; Webster, I, 154-5, whence the quotation is taken.

Prussia and Austria must be hurt by it. If the Powers adhere to it, they ought to exact from Russia a frank adhesion to 24 Articles." [2] But Palmerston prevailed on the Conference to accept the Russian ratification in London and thus he kept them together.

Still, it was little more than sitting together, since the whole Conference, except the Foreign Secretary, did not care for Belgium; moreover, as he knew, the Russian plenipotentiaries "want to recommend themselves to their Emperor and to win orders and recompenses and make atonement for the crime of having signed the Treaty of November".[3] Something had to be done to make the proceedings of the Conference more effective. In the summer of 1832 Lord Durham was sent to St. Petersburg with instructions to try to have Nicholas agree to the necessity of coercive measures with regard to Holland. Should he not concur, at least he ought not to oppose them. Since, however, Durham's departure from London coincided with a wave of public indignation in Britain on account of Russian measures in Poland and the envoy himself was considered to be a partisan of the Polish cause, his mission was acclaimed by "The Times" as an intervention on behalf of Poland. In fact, Durham's mission was far more a personal affair within the Whig Party, as Grey wanted to offer him some distraction after his daughter's death. The mission itself did not affect Belgium and Poland very much and, because of Durham's character, its influence was felt in Whig relations rather than in international politics.

For the rest, linking up Polish and Belgian affairs no longer seemed possible; Nicholas did not admit any right of interference with regard to Poland on the part of the Western Powers, and the British Ministers, in spite of all the parliamentary difficulties created by the news from Poland,[4] were well aware that their protests to Russia lacked strength. Nevertheless, such protests continued to be made throughout the better part of the XIX century, especially in 1836, 1846 and 1863, though with no force to support them they

[2] Leopold to Le Hon, 4 Apr., *Memoirs of Leopold I,* I, 253.

[3] Webster, I, 155.

[4] Thus Palmerston wrote to Melbourne on 10 April, 1836, when the crisis around the Republic of Cracow again drew the attention of Parliament to Poland: "I remember that in the former Polish affair Lord Grey insisted upon much softening down the dispatches which I had drafted, alleging that they would give offence to the Emperor, and afterwards, when Cutlar Ferguson moved for their production [in Parliament] Lord Grey desired that they should not be given because he said he felt that we had not said enough." Panshanger MSS.

carried little weight; there had been no Polish army on the Vistula since 1831.

Belgian affairs which had progressed rapidly in 1830 and in 1831 became the subject of a slow process of negotiations. After the breakdown of the Polish movement, Matuszewic, the second Russian plenipotentiary, became "la bête noire" of the Conference for the British Ministers;[5] the other Eastern representatives too took a much firmer line in favour of Holland during the negotiations that followed than they had previously done. During the early months of 1832 considerable delay in ratification of the Treaty gave rise to some false interpretations, but no other encouragement was afforded to the King of the Netherlands; his representative in London, Van Zuylen, had already in November, 1831, predicted such a development and attributed the absence of such assistance as the King would desire to the exhaustion of Russia caused by the Polish war[6]; indeed, the three Eastern Powers did not object to the expedition of a French force which, at the end of 1832, besieged and captured the Antwerp citadel, held until then by the Dutch army.

After the Convention of 21st May, 1833, which provided for a *modus vivendi* without the recognition of King Leopold by The Hague, the Conference adjourned *sine die* and the affairs of the Low Countries were left to take their own course. The final settlement took place six years later. In the negotiations, renewed about a year earlier, the Belgians were not allowed to draw any advantage from the fact that during the intervening years they had held portions of territory assigned in 1831 to Holland; they were compelled to abandon them. It remains to be noted that public indignation in Belgium at the prospect of giving up what was considered an integral part of the national territory rose to such a height that the Government seriously reckoned with the possibility of being forced into war, and hastily made such improvements in its armed forces as were possible within so short a time; in these measures the appointment of a number of Polish officers, who went into exile after 1831, was included and among them was General Skrzynecki, the Polish Commander-in-Chief in 1831. The withdrawal of the Austrian and Prussian Ministers from

[5] Neumann to Metternich, 26 Oct., 1832, quoted by de Lannoy, p. 55.
[6] Van Zuylen to Van der Hoop, 25 Nov., 1831, Ged., IV, 609.

Brussels – Russia had yet no representative there – made it impossible for Skrzynecki to continue, and he offered his resignation.

It is sufficiently established that the British Government was mainly responsible for the recognition of the Orleans Monarchy in Europe and for arranging an assembly of the Great Five to discuss the affairs of the Low Countries. The views originating in London and Paris supplied the basis for the existence of the Belgian State, in spite of any differences on this account between the two governments. The question necessarily arises why the other three Powers, Austria, Prussia and Russia, parties to the arrangement in virtue of which the united Kingdom of the Netherlands had been created, exercised so comparatively little influence in bringing about a new settlement in the Low Countries; and the answer is that they were more than handicapped by the Polish revolution. It has been at times underlined by historians that the outbreak of the revolution in Poland, which considerably embarrassed the three Eastern Governments, enabled the Governments of France and Britain to obtain the London Conference's agreement on the independence of Belgium. This opinion slightly exaggerated the importance of the Protocol of 20th December, in which the term "indépendance future" with regard to Belgium was employed for the first time. An attempt has been made here also to stress the haste with which this measure was adopted and the fear of the French Government that it still might be undone by an untimely failure of the Polish movement. Hence the attempts of the French Government to draw the attention of Britain to Poland. Hence, too, its aquiescence in the candidature of the Prince of Orange for the throne of Belgium, though its success would tend to maintain the link between the two parts of the former kingdom of William I, contrary to the wishes of France. Lastly, efforts made from Paris to bring the Belgians to agree to the "Bases de séparation" are to be ascribed to the same cause; these efforts continued to be made until April, 1831.

For the rest, the influence of the events in Poland upon the course of the Belgian policy of the Eastern Powers is better known, chiefly because of the work done on the subject by German historians. One of the main questions which I have tried to answer was whether the obvious divergence of views between Britain and France in the affairs of the Low Countries was also responsible for their policy, and especially for the policy of Britain, towards the revolution in Poland. More attention is given here to this problem as, unlike the Belgian affairs, it has so far only been examined to an inconsiderable degree.

The outbreak of the revolution in the Kindom of Poland, and later, in other parts of the former Polish State, could not fail to attract the attention of the British Government; Poland was one of the key problems at the Congress of Vienna. Moreover, there was widespread sympathy for the Polish movement in Britain also, though this was not comparable to the excitement prevailing in this respect in the French capital. Little as the British Government was in a position to afford even a minor quarrel with Russia over Poland, some representations concerning Polish affairs were nevertheless made in St. Petersburg by the British Ambassador, and they seemed there to be the result of a suspected common policy of Britain and France in this respect. In April these coincided, too, with bad news for Russia from the theatre of military operations. The interest of France in Poland was not unexpected by the three absolutist governments; that of Britain caused them some surprise, especially as they were aware that Britain was opposed to France over the Belgian question. They realized, however, that, on the whole, London was closer to Paris than to the other Courts, or at least likely to be so, and therefore in order to prevent the British Government drawing too close to the Palais Royal at the time of the Belgian crisis in June, 1831, they had given way at the expense of William I. Yet, only a few weeks later, after Belgium appeared to be appeased, and whilst the French invitation to Britain to join in a diplomatic intervention in St. Petersburg in favour of the Poles was being discussed in London, the Prussian observer there was convinced that war because of Poland was inevitable.[7] In July, too, diplomatic circles in St. Petersburg smelled intervention.

In fact, the Poles had received no promises from the British Ministers on which to base any hope for assistance except Palmerston's statement that he considered Poland lost to Nicholas. This statement seems to have made considerable impression upon the Eastern Governments; its echoes circulated there during August,

[7] Bülow to Frederick William III, 16 July, 1831, quoted by F. Richter, *Das europäische Problem der preussischen Staatspolitik und die revolutionäre Krisis von 1830 bis 1832*, 1933, p. 109.

when the Russian army was about to lay siege to Warsaw. The possible effects of such an impression, even if caused by an Anglo-French *démarche*, were not appreciated in London, although it does not seem that Palmerston made his statement without a purpose; if what he told Walewski was meant to bring about a Polish intervention in Brussels for the sake of the XVIII Articles, the words spoken to Matuszewic were destined for St. Petersburg. At this time, however, the majority of the Cabinet believed that an offer of mediation could only result in an open breach with Russia. As Lord Holland put it, when writing to Granville: "You all seem to me to underrate the authority with which Great Britain and France when acting conjointly can spreak to other Powers, and Russian particular." [8]

The French initiative was therefore rejected. It is clear that the still unstable state of Belgium contributed to furnish ground for this decision; to what extent, it is difficult to assess. The official reply spoke of the loyal attitude of Russia in the Belgian negotiations; it has been shown that Palmerston disagreed with his colleagues and did not want to have this motive included in the document sent to Paris.

How far was St. Petersburg prepared to tolerate interference? The official attitude left no doubt: Poland and especially the constitution of the Kingdom as granted by the Emperor Alexander, was Nicholas's concern alone. It seems that in the beginning Nicholas might have been willing to make some political declaration, if the Poles were to submit; the first audience of Mortemart, as well as letters to the Grand Duke Constantine at a later date provide some evidence for this attitude.[9] But it was fairly unlikely that the Poles would adopt such a course. In May and June the military position of Russia deteriorated and Nicholas had second thoughts on the ultimate fate of his Polish possessions. He considered a partition of the Kingdom of Poland as a means which, apart from its political significance for Russia herself, might bring about the armed assistance of other interested Powers, Prussia in particular.

At that time, however, there was no likelihood of moving the Prussian army across its Eastern frontiers: not only was it obvious in Berlin that France would in such a case try to seize the Rhine-

[8] Holland to Granville, 19 July, Granv MSS. See Appendix XI.

[9] Cf. above, Chapter II, p. 94, and Chapter III, p. 130.

land, but as the activity of the Polish agents in Paris and London grew, the Prussian Government found itself confronted with protests regarding its neutrality in the Polish war only on account of supplies afforded to the Russian army. The special Ambassador of France, Flahaut, was convinced that public opinion in Germany alone would suffice for the Berlin Government to keep clear of the Polish war.

There was no likelihood of Austrian assistance either. Though Palmerston made much of Dwernicki's case[10] that Austria favoured Nicholas, Vienna was far from being prepared to compromise herself on account of Poland: on the one hand, Italian affairs affected her much more than the others; moreover many persons in the Court and Government circles were sympathetic to Poland, among them the Governor of the Polish province of Austria, Galicia; in addition the Hapsburg Empire was hardly able to bear the strain of military expenses already upon it. Besides, in Metternich's view, the Polish revolution led by the aristocracy, headed by Prince Czartoryski, could not be blamed for carbonarism. But if there was no certainty that the Western Powers had made up their minds to intervene decisively in Polish affairs, the Chancellor could not be expected to jeopardize his relations with Russia and Prussia by showing sympathy towards the Poles. Hence the obvious inconsistency of the views of Metternich, as ascribed to him by the British Ambassador, especially in his reports during July, 1831. Nicholas accordingly made no request for assistance to Austria; but he declared to the Austrian Ambassador that rather than submit to foreign interference, he would abandon Poland altogether – by which he meant the "Congres"-Kingdom only – as if he knew that in many a project of intervention the participation of Vienna was envisaged.

There is a document showing that the abandonment of Poland seriously occupied Nicholas' mind.

In May, when the military position appeared rather gloomy, Nicholas wrote a personal memorandum concerning Poland. The Kingdom had to be conquered, he noted: the dignity of Russia could not remain intact after such blows as she had already suffered; but there were really no advantages in keeping Poland. This Kingdom should therefore be partitioned between

[10] Cf. Chapter IV, above, p. 152 and 180.

Prussia, Austria and Russia; that part which would be Nicholas' should keep the name of Poland in order that conflicts be excluded which might follow if one of the other parts assumed that title.[11]

The memorandum showed confusion in the Emperor's mind since the carrying out of the project would require the co-operation of the other two Powers, and yet the condition made that the name of Poland should remain with the Russian portion, lest it were used by one of them, indicated that they were not to be trusted. In any case, when Orlov brought this plan with him to Berlin in June, he obtained no official reply and indeed the Prussian Government did not seem to have considered it seriously. The importance of Nicholas' project, however, lies in that part of the document in which he put forward his reasons why it should be accepted: not only was the possession of Poland useless, but dangerous, even menacing as well: "Only to glance at the map" wrote Nicholas "is to be horrified at the sight of the Polish boundaries: they extend to the Oder, whilst at the sides they reach behind the Nieman and the Bug in order to touch the Baltic near Polangen and the Black Sea by the estuary of the Danube."

This opinion, in itself explicit enough considering that it was destined for foreign eyes as well, seems to leave a door open through which to introduce Talleyrand's advice of the previous December that the Imperial Government might, in time, be brought to yield "à des démarches habilement combinées." [12] Nesselrode's influence, just as in the autumn of 1830, might perhaps have been used for this purpose. However, difficulties in the way of such *démarches* as suggested by the French Ambassador in London were considerable. To begin with, any diplomatic intervention was very seriously handicapped by the dethronement of Nicholas by the Polish Diet, in January, 1831, and still more, by the fact that in the course of the revolution the Poles insisted on claiming not only the independence of the Kingdom of Poland, but also that of the Polish provinces incorporated in Russia as well. Once the principle of non-intervention seemed in 1830 to have been accepted, there was very little room left for meddling in the affairs of Poland which, moreover, were really those regarding all the former Polish territories; it necessarily would lead to some sort of breach with Russia.

[11] Schiemann, *Geschichte Russlands,* III, 120.
[12] Cf. Chapter II, above, p. 85.

Even were it taken into account that the Treaty of Vienna provided for special treatment of those territories by all the partitioning Powers, the act of the Congress at the same time stated that the "Congress"-Kingdom – let alone the other provinces – was "irrevocably tied" to Russia. But at the time in July, when the intervention was likely to become a reality, the Polish Government might more easily have been brought to a compromise than half a year earlier.

The British Government could not, and would not, face the likelihood of a serious breach with the Eastern Powers; it was already, as Talleyrand put it, overburdened with two important questions, Belgium and the Reform. Belgium would in any case remain a problem even if frank co-operation existed between Paris and London. But neither were the Eastern Powers in a position to risk a war, whether against both the Western Powers, or against France alone. It took time before the two Powers became aware of this fact; it may well be assumed that the interest of the French Government in the Polish question during the winter months of 1830/1831 was primarily directed by consideration for its own safety. France herself did not need to be afraid of a war: with feelings all over Europe stirred up by the "trois glorieux" in Paris, Prussia and Austria realized that war against France was only too likely to be supported by revolutionary movements in other countries, possibly in Germany as well. But the Orleanist régime, and especially the "Bourgeois-King" himself were aware that such a war would bring a speedy end to their weak authority; but that did not need to be a reason to prevent them from using threats of such a war in their relations with Berlin, Vienna and St. Petersburg[13]; more than that, the less faith the Eastern Governments placed in the sincerity of the pacific assurances of the French Government, the more likely were they to believe in such an eventuality even without the threats actually being held out to them.

As is known, all that the Palais Royal did for the Poles was literally forced upon them by the public opinion.[14] Not only did the Polish mission in Paris have to be "managed", but Polish radicals staying in France associated themselves only too easily with extreme elements there; Palmerston complained about it to Walewski. That explains why the invitation to Britain to join in an interven-

[13] Cf. above, p. 96, reference to the alleged interview of Sebastiani and Pozzo.
[14] Cf. especially Dutkiewicz, *Poland and France in 1831*, *passim*.

tion on behalf of the Poles was made in July; at the same time the French Government badly needed some popular success since Leopold's arrival in Belgium had put an end to whatever plans were entertained in various circles in Paris with regard to the North-Eastern frontiers of France.

The attitude of the British Government was much more frank. The British Ambassador in St. Petersburg received instructions on the subject of the "Polish" stipulations of the Treaty of Vienna not only before the arrival of Mortemart but even before the actual beginning of hostilities on the Vistula, as though the British Government recognized the violation of the constitution which was put forward by the Poles as the cause of the rising. Until April, however, no more was done since the military operations did not warrant the hope that the Polish movement would maintain itself against the Russian onslaught. From April onwards London seemed to be prepared to act only as the result of a decisive Polish success – at least the continuance of the movement throughout the summer – since this would imply the prolongation of the war in Poland beyond 1831 and cause unavoidable repercussions elsewhere.

Unfortunately, the very important interview of Walewski with Palmerston in June created a false impression for the first time. Without saying that he would intervene Palmerston stated that Belgian affairs prevented such an intervention. All the more disappointing, therefore, appeared to be the British refusal to join in an intervention, proposed in July, at a time when Belgium had obtained what seemed to be an appropriate solution, and at a time when public opinion in Britain had become a factor underlining the need of a more active policy.

The state of affairs in Poland was by no means satisfactory for the insurgents; but it is hardly to be assumed that this contributed considerably to the British Government's decision, as it no doubt might have done if London had had accurate accounts; as it was, news from the theatre of military operations was scarce and hardly reflected the real situation, while diplomatic reports from the Russian capital, which in July concentrated on the effects of the cholera epidemic, could only strengthen the impression of the weakness of the Imperial Government. As previously stated, it is difficult to give an opinion on the part played by Belgian affairs in bringing the British Government to refuse the French overture. It is clear, however, that London's reference to the Belgian policy of

Russia neither approved of it nor constituted a declaration against the Poles but was pointed against the Belgian policy of France. As for Poland, the British note distinctly stated that the Government would try to intervene in such a way as would be consistent with its relations with St. Petersburg.

If the possibility of further complications in the Belgian business was sufficient to make the British Government abstain from straining its relations with the Eastern Powers, it need not be further explained that the renewal of hostilities in the Low Countries and the subsequent entry of a French army into Belgium put further active interest on the part of London in Polish affairs entirely out of the question. Suggestions made by Talleyrand with regard to linking the French occupation of Belgium and the Polish war missed any prospect of realization; and if they were brought forward at all, they only provide evidence that the French Ambassador's "idée favorite" – the partition of Belgium – obscured his better knowledge of Britain. The British Government, if compelled to agree to such a solution, as, according to Palmerston view in June, they might have to under certain circumstances, would do it only as the very last concession; Polish affairs hardly would be expected to play a rôle at all in such a bargain.

It may be observed that the Belgian inability to deal with the Dutch attack provided quite a sound argument for the French to keep their army in Belgium. The British Government was unable to expose France's bad faith except by such slips as the French Ministers might make in their relations with their foreign colleagues.

The necessary conclusion appears to be that uneasy relations between Britain and France were responsible for the absence of any intervention in Poland, an intervention which was loudly demanded by public opinion, was likely to bring some results for Poland and – which unfortunately was not sufficiently appreciated – was one that involved no great risks. It is beyond the scope of this work to analyse how far the situation in Poland could be adapted to meet such an intervention; it is sufficient to state that the news of it would strengthen the moderate elements, perhaps to the extent of making the August riots impossible.

Finally, Grey's remark about sending a fleet to the Baltic by France must be mentioned. Evidence is lacking to explain the absence of any such step. The French Government apparently did not dare to intervene on its own; as explained above, by so doing it would probably undermine its own existence, since a distant foreign expedition, with the possibility of involving hostilities against all the three Eastern Powers, might necessitate appealing for popular support. The present study hardly gives an answer to the question why there was no hint from London to Paris regarding a naval expedition on behalf of Poland. Such a hint would dismiss the fear of a large scale war. A similar hint, Palmerston's declaration that Britain would not support the Eastern Powers against France if they intervened actively in Poland, obviously greatly contributed in making them abstain from offering military assistance to Russia. The appearance of a French fleet in the Baltic would undoubtedly have brought Prussia and perhaps Austria to interpose their mediation, not only in order to avoid a possible general war, but even to prevent the Poles obtaining supplies via the sea route; the French Government would have been equally eager not to get involved in actual hostilities. The effect of such mediation cannot be dwelt upon here; this would of course also be influenced by the last stage of the military operations in Poland, a subject which goes beyond the aim of the present study.

APPENDIX I

Count Matuszewic to Count Nesselrode, 8th October, 1830

Algemeen Rijksarchief, The Hague. Ref. Aanw. 1918/xi, no. 30. Enclosed to the dispatch of Baron Van Heeckeren, dated St. Petersburg 1/13 November, 1830, under the title: "Extrait d'une dépêche du Cte de Matouchewic du 26.IX/8.X.1830, no. 207."

Le système que vient d'adopter le cabinet de St. James me paraît se réduire aux termes suivans:

Le cabinet [1] de St. James croit ne pouvoir refuser au roi des Pays Bas le conseil et le secours que ce monarque réclame aujourd'hui de la part de l'Angleterre, de la Russie, de l'Autriche et de la Prusse.

Il regarde comme dangereux d'accorder immédiatement des secours militaires, ou l'impossibilité de rendre ces secours assez considérables pour en assurer l'effet.

Il juge indispensable à sa position et à ses nécessités parlementaires comme au bien général, d'épuiser toutes les chances de paix, en essayant d'associer le gouvernement Français aux délibérations qui auront pour le but le nouveau mode de coëxistence de la Belgique avec la Hollande.

Dans cet arrangement, il est décidé à n'admettre ni une séparation de Dynastie ni aucune combinaison qui placerait à la disposition de la France, la barrière de forteresses érigée contre elle.

Si la France veut soutenir maintenant le principe qui excluerait à l'avenir toute intervention militaire, toute mesure coërcitive, l'intention du cabinet de St. James est, non de souscrire à cette demande, ou de laisser croire que jamais il y souscrira, mais d'ajourner la discussion, en la déclarant prématurée.

Si cette demande se reproduit, quand il s'agira de mettre à exécution les conditions nouvelles arrêtées avec le Roi des Pays Bas,

[1] The following paragraphs, translated into English, are included in the dispatch of Lord Heytesbury, dated St. Petersburg, 22 October, 1830, F.O. Russia.

le cabinet de St. James opine à passer outre, alors, et à n'en tenir aucun compte, en cas que les Belges refusassent à se soumettre aux décisions de l'alliance.[2]

Il est évident que les premières réponses du cabinet Français et surtout les termes dans lesquels sera ajournée la question des mesures coërcitives, vont résoudre le grand problème qui nous occupe.

Si le gouvernement Français consent à l'ajournement de cette question, sans emporter le principe qu'il soutient jusqu'à cette heure, il aura fait évidemment un pas rétrograde, il se sera rapproché des doctrines des 4 cours et il aura de la peine à faire revivre plus tard ses étranges prétentions.

S'il veut dès à présent que l'Angleterre s'y soumette, j'aime à espérer que M. de Talleyrand trouvera dans le caractère vigoureux du premier ministre anglais toute l'opposition que j'attends de ses maximes, de la haine pour les révolutions et de la gloire dont il s'est couvert en les combattant.

[2] Here ends the part included in Heytesbury's dispatch, cf. note 1.

APPENDIX II

Nicholas I to his sister, Anna, Princess of Orange, 16th December, 1830

"Copie d'une lettre autographe de l'Empereur à la Princesse d'Orange." Lieven MSS at the British Museum (Add MSS 47398).

St. Pétersbourg, le 4/16 Xbre 1830

Il part un Courrier dans quelques heures...[3] et je veux néanmoins te dire que j'ai reçu tes deux lettres par Trubetzkoï et par Courrier, je ne puis assez te remercier et te dire combien j'apprécie l'amitié que tu me témoignes. Voilà donc que la révolution et tout ce qui s'en suit nous a atteints aussi comme je le prévoyais. Tu sais déjà tout [ce] qui se passe en Pologne, il y a demain 7 jours que je ne sais rien de Constantin, on le dit parti avec les troupes russes pour rentrer dans l'Empire. Tu peux penser la besogne que j'ai sur les bras, mais je me rends avec confiance à la volonté de Dieu qui m'a tiré de position plus difficile que celle-ci. J'espère qu'il nous protégera et me fera m'acquitter de mes devoirs pénibles qu'il m'inspirait. Sans cet aide tout puissant je ne saurais rien faire mais avec lui j'envisage le danger de sang froid. Nos troupes marchent; dans quelques semaines je devrai les rejoindre. Le reste est écrit là haut et je m'y soumets d'avance. Mes principes sont irrévocables et je périrai plutôt que transiger avec la révolution. Tu conçois que dans des momens pareils l'on ne peut qu'être laconique et tu ne m'en voudrais pas. Mille & mille remercimens pour tous les voeux que tu fais pour moi à l'occasion de ma fête et pour le charmant achat.

Si Guillaume est près de toi, embrasse le moi, ainsi que les autres. Adieu, pense à nous et que Dieu te conserve avec les tiens.

à toi pour la vie
ton dévoué frère et ami
Nicolas

[3] Two words illegible.

APPENDIX III

Protocol No. 7 of the London Conference on Belgian Affairs

A. DRAFT MADE BY TALLEYRAND

An enclosure to the dispatch of the Netherlands plenipotentiaries A. R. Falck and Baron H. van Zuylen van Nyevelt, of 28th December, 1830 (Algemeen Rijksarchief, The Hague, Ref. A. III, no. 9.)

Les Puissances signataires du Traité de Vienne avaient à remplir les unes envers les autres et envers toutes celles dont l'Europe Chrétienne se compose la Ière et la plus grande des obligations – celle de favoriser dans tous les Etats avec le progrès de la civilisation l'accroissement et la durée du bien-être dont elle est la source en assurant la plus longue paix pour l'établissement d'un équilibre, si non parfait, du moins aussi réel que pouvait permettre l'état respectif des possessions que 22 années de guerre avaient amené.

L'union des nations Belge et Hollandaise en un seul peuple sous le régime libre et sous le Gouvernement de la Maison de Nassau avait paru et dû nécessairement paraître une condition principale et pour ainsi dire nécessaire à l'équilibre à établir et cette union fut sanctionnée dans l'intime persuasion qu'elle serait également utile et bientôt également agréable aux deux nations; car rien alors n'avait fait...[4] et personne ne pouvait imaginer qu'il y eut entre elles une incompatibilité que le tems ne pouvait vaincre et telle que l'ont révélée les événemens des trois derniers mois.

Tout en regrettant que cette union n'ait mieux répondu aux espérances dans lesquelles elle avait été conçue, les Puissances signataires du Traité de Vienne se regardent comme entièrement libérées, par nature même de l'obstacle imprévu contre lequel elle

[4] One word illegible.

n'a pu se soulever, de toutes obligations résultans de la combinaison qui avait d'abord été adoptée.

Mais libres du devoir qu'elles s'étaient imposées elles s'en sont trouvées un autre qui est devenu plus impérieux par cette circonstance même.

Gardiennes [5] du repos de l'Europe les Puissances signataires du Traité de Vienne n'ont voulu employer vis-à-vis d'autres voyes que celle d'une intervention toute pacifique, ni d'autres forces, que celle de la raison dont elles ne se doutaient pas qu'elle ne fut tout puissant sur l'esprit des états investis de la confiance des populations Belges et sur ces populations elles-mêmes.

Leur attente n'a pas été trompée. Les Belges ont senti que situés au coeur même de l'Europe leur nouvelle existence devait porter à la grande famille Européenne de la sécurité et non de l'inquiétude. Le principe monarchique a été conservé. Déférant ensuite avec empressement à l'armistice qui a été proposé par la conférence de Londres ils ont mérité qu'on eût en eux actes de confiance que les Ambassadeurs des cinq puissances déclarassent à l'Europe qu'elle avait acquis un nouveau royaume et que les Belges étaient indépendants.

Art. 1. Les cinq grandes puissances réunies à Londres accèdent aux voeux des Belges et les reconnaissent comme un peuple formant un état séparé de tout autre et indépendant.

Art. 2. La Belgique étant appelée à prendre rang parmi les Etats Européens devra, dans l'organisation définitive qu'elle donnera à son gouvernement monarchique, avoir ainsi en vue de n'inquiéter aucune des puissances au milieu desquelles elle va se trouver placée.

Art. 3. Tout ce qui tient aux difficultés relatives à la séparation de la Belgique et de la Hollande sera réglé par des Commissaires respectivement nommés, avec recours, en cas de différence, aux commissaires des cinq puissances médiatrices.

Art. 4. Il est entendu qu'il n'est par la présente reconnaissance rien statué [6] qui puisse affecter les droits que la Confédération Germanique et le Roi de Hollande peuvent avoir sur le Grand Duché de Luxembourg.

[5] The remaining part of this document is printed in a footnote to p. 84 of Fl. de Lannoy's *Histoire diplomatique de l'indépendance belge,* 1930.

[6] In print "établi".

B. PROTOCOL SIGNED ON 20TH DECEMBER, 1830

From *Recueil des piéces diplomatiques relatives à la séparation de la Belgique d'avec la Hollande* [by J. G. Verstolk van Soelen], La Haye 1832, I, 44-46.

Les plénipotentiaires des cinq cours ayant reçu l'adhésion formelle du gouvernement belge à l'armistice qui lui avait été proposé et que le Roi des Pays Bas a aussi accepté, et la conférence ayant ainsi, en arrêtant l'effusion de sang, accompli sa première tâche, qu'elle s'était imposée, les plénipotentiaires se sont réunis pour délibérer sur les mesures ultérieures à prendre dans le but de rémédier au dérangement que les troubles survenus en Belgique ont apporté dans le système établi par les traités de 1814 et 1815.

En formant, par les traités en question, l'union de la Belgique avec la Hollande, les puissances, signataires de ces mêmes traités, et dont les plénipotentiaires sont assemblés dans ce moment, avaient eu pour le but de fonder un juste équilibre en Europe et d'assurer le maintien de la paix générale.

Les événemens des quatre derniers mois ont malheureusement démontré que "cet amalgame parfait et complet que les puissances voulaient opérer entre ces ceux pays", n'avait pas été obtenu; qu'il serait désormais impossible à effectuer; qu'ainsi l'objet même de l'union de la Belgique avec la Hollande se trouve détruit, et que dès lors il devient indispensable de recourir à d'autres arrangemens pour accomplir les intentions, à l'exécution desquelles cette union devait servir de moyen.

Unie à la Hollande, et faisant partie intégrante du Royaume des Pays Bas, la Belgique avait à remplir sa part des devoirs européens de ce royaume, et des obligations que les traités lui avaient fait contracter envers les autres puissances. Sa séparation d'avec la Hollande ne saurait la libérer de cette part de ses devoirs et de ses obligations.

La conférence s'occupera conséquemment de discuter et de concerter les nouveaux arrangemens des plus propres à combiner l'indépendance future de la Belgique avec les stipulations des traités, avec les intérêts et la sécurité des autres puissances et avec la conservation de l'équilibre européen. A cet effet la conférence, tout en continuant ses négociations avec les plénipotentiaires de S.M. le Roi des Pays Bas, engagera le gouvernement provisoire de la Belgique à envoyer à Londres le plutôt possible des commissaires munis des instructions et des pouvoirs assez amples, pour être consultés et entendus sur tout ce qui pourra faciliter l'adoption définitive des arrangemens dont il a été fait mention plus haut.

Ces arrangemens ne pourront affecter en rien les droits que le Roi des Pays Bas et la Confédération Germanique exercent sur le Grand Duché de Luxembourg.

Les plénipotentiaires des cinq cours sont convenus que le présent protocole serait communiqué au plénipotentiaire de S.M. le Roi des Pays Bas, et envoyé en copie à Lord Ponsonby et à Monsieur Bresson, moyennant la lettre ci-jointe, dont ils donneront connaissance au gouvernement provisoire de la Belgique.

APPENDIX IV

Viscount Palmerston to Lord Ponsonby, Foreign Office, January 4th, 1831

Broadlands Archives

Private.

My dear Lord Ponsonby,

We had a Conference today at which the Dutch Plenipotentiaries communicated a long protest by the King against our 7th Protocol, requesting that this document should be annexed to a Protocol of our proceeding.

But the document is so worded as to make it impossible for us to comply with this request, unless it is materially altered. They then stated that the King proposes to take off the Scheldt blockade on the 20th of this month, hoping that by that time the material points of the arrangement for the separation of the two countries may be settled and they stated his notions of such an arrangement to be as follows.

1. That the territory of Holland should be that which she possessed in 1794, before the conquest by France, but that exchanges should be made between her and Belgium so as to get rid of some of the...[7] and to give to Holland a military road to Maestricht.

2. That the debt should be divided, each party taking which belonged to them before the union and their just proportion of that incurred since the union.

3. That if Belgium will further take upon herself a larger portion of the debt, the amount of which increase shall be hereafter fixed, the King will give to Belgium the same privileges of commercial intercourse with the Dutch colonies which is enjoyed by his own Dutch

[7] One word illegible.

subjects and still further allow free navigation of the Scheldt to Antwerp.

4. These heads of arrangement seem to be fair enough and we are going to consider them, and hope in the course of next week to settle something about them.

Van de Weyer and his colleagues have powers to discuss only and not to conclude, we must therefore settle the territorial question for the two parties and when that matter and the fortress question are arranged, the Sovereign may be chosen forthwith.

Van de Weyer told me today that Cobourg was going down and Nemours getting up.

I told him that the Allies never would consent to Nemours. But said he, what if the Congress were to chuse [sic] him? Isaid that it would then become a question between us and France. That we should expect France to refuse her assent, or that war between France and the rest of Europe might be the necessary consequence.

It may be well for you to hold the same language, and with regard to the chance of France hazarding such a war it should be remembered that the events of the last ten days at Paris [8] have given the King and his Government great additional powers of restraining popular feeling, by having much strengthened their authority.

That Russia will probably make some amicable arrangement with Poland and be ready therefore to act in case of need, and that our disturbances in England have been pretty well mastered, and that Ireland will I trust be kept quiet. That the ballot and training of the militia will give us 50,000 available men and that consequently the forces with which France would have to contend in case of a war about Belgium are more formidable that they were a fortnight ago. Talleyrand showed his cloven foot in a conversation I had with him the day before yesterday. I said an arrangement had occurred to me which might or might not be practicable but which if it could be accomplished might remove many difficulties – "Why should not the Belgians take the Prince of Orange on condition that the King made over to him his rights on Luxembourg"?

Talleyrand looked grave, said he doubted whether his Government would agree, that Luxembourg was too near to them and ran

[8] Reference is made to the process of the Ministers of Charles X. Louis Philippe withstood popular pressure which demanded passing the death sentence on them; the King went so far as to remove Lafayette from his post of Commander of the National Guard.

into their frontier. – "Not more said I, as attached to independent Belgium than as belonging to the United Netherlands", "Aye said he, let us have no town, then our frontier is open and exposed;" "why, said he, could not some arrangement be made *by which Luxembourg might be given to France"?* I told him at once that never could be, we could not agree to it, we wish to be well with France, but it could not remain so, were France to aggrandize herself in that direction, that it is an English interest that Belgium should be independent, and for that same reason France could not have Luxembourg.

This is the first time he had let out any distinct proposition for adding to the French territory, although one saw clearly that his proposal of the Saxon arrangement [9] would, if agreed to, have led to some proposition or other of this kind.

I conceive the Prince of Orange to be completely out of the Question, and that Leopold is the best candidate, but Russia has positively forbidden his Minister here to consent to any Sovereign not a Nassau; Prussia has left her full discretion by giving *no instructions.*

If we settle other matters as I hope, the Sovereign may be managed – I do not think that you can mention the Dutch proposition as one coming from the King, but we might, if you think it would be useful, hint that such an arrangement is what the Conference is aiming at.

We hear that when it was supposed that the Conference had decided that Luxembourg was to be immediately given up the deputies from the Duchy were on the point of retiring from the Congress, not to create unnecessary difficulties. Do you believe this to be true? It is quite certain, however, that Luxembourg is settled by too many treaties to make it possible to give it to Belgium, except by the consent of all the parties interested.

My dear Lord Ponsonby,
Yours very sincerely,
Palmerston

[9] Talleyrand's proposal made to Bülow whereby Prussia would acquire Saxony, referred to above, pp. 79-80. Cf. W. von Franqué, *Luxemburg, die belgische Revolution und die Mächte,* 1933, p. 261.

APPENDIX V

Viscount Palmerston to Sir Charles Bagot, 12th January, 1831

Bagot MSS. Levens Hall, Westmorland.

F.O. 12 Jany 1831
R. Ja^{y} 15 at 3 PM by Mess. Tricker

My dear Bagot,

We have pretty authentic information that a considerable change of public opinion has lately taken place in Belgium and that the Orange Party begins to shoot ahead. The Prince has in consequence been advised to prepare a sort of proclamation if it can be so called and to send it to some of his confidential friends at Brussels that they may spread it as far and as soon as they can. I send you a copy. The Prince has also written to the King a letter sent off last night [10] to tell the King that he may probably have the Crown of Belgium offered to him and to express his hope that in such an event the King will not oppose any obstacle to his accepting it. Such an arrangement holding out the best prospect of a speedy settlement of Belgium and the only chance of retaining the country for the House of Nassau. The whole of this matter and proceeding have been kept strictly secret, in so much that no communication has been made about it either to Falck or Zuylen. But the Conference have requested me to write confidentially to you to beg you to obtain without delay a private audience of the King for the purpose of expressing to His Majesty the earnest hope of the Five Powers that if there should appear to be any chance of the election of the Prince of Orange to the throne of Belgium, His Majesty may be induced to afford every facility in his power to such an arrangement. The King may at one time have entertained a belief that it was still possible that the two crowns might be united upon his own head but the Conference are persuaded that such an arrangement has

[10] This letter is printed in F.C. Gerretson, *Muiterij en scheuring*, 1936, I, 96.

long become impracticable and must see that by pursuing in impossibility He may lose to his son and his throne that possibility which if it be not grasped at the moment may rapidly become also an impossibility.

The Conference wish you moreover in the strongest manner to impress upon His Majesty that nothing would afford to the Prince of Orange a better Prospect of success than his being authorized to declare on the part of the King that the Grand Duchy of Luxembourg should either now or hereafter be made over to him; the importance of that Duchy or Belgium is sogreat and so deeply felt that the certainty of having it attached to Belgium provided the Prince of Orange were elected would of itself almost secure H.R.H.'s election. To Holland the Duchy is of comparatively trifling importance, being cut off from it by the interposition of Belgium, and to the King himself the revenue to be derived from it cannot be considerable and might be made subject of a pecuniary arrangement with Belgium.

Of course I do not propose that Luxembourg should cease to form a part of the Germanic Confederation to which on the contrary it is most important that is should continue to be attached.

It is most essential that is should not be known that the Conference had anything to do with the steps taken by the P. of Orange, because the mere knowledge of that fact might perhaps lead the Belgians to reject him and this is the reason of the great secrecy we recommend. But if the thing takes in Belgium and you find that any considerable party declare in the Prince's favor, some public act by the King relinquishing his rights in favour of his son; and promising friendship and good neighbourhood and commercial facilities to Belgium would probably be very useful. The King is I presume not in the best of humours with respect to our last protocol, but that does not signify, it *is all for his own good* and our present suggestion ought to convince him of this, as it certainly is a proof that all the Powers are sincere in wishing well to the House of Nassau.

Zuylen hinted to me that the Scheldt might be opened to all vessels bearing a recognized flag, I told him it was not worth while to [11] ...about the question of flag; he had just told me that when at sea a Dutch ship of war met a Belgian flag, the Dutch man shut

[11] One word illegible.

his eyes, I told him they had better do the same in their coast batteries on the Scheldt.

I forgot to request you to give notice to our consuls and merchants that the Scheldt would be opened on the 20th of this month, pray do so.

My dear Bagot,
Yrs sincerely,
Palmerston

You may tell them that the insurrection of the clods in the South of England is over, as it is said, for half a century to come, and that the Irish Rebellion is put off sine die.

APPENDIX VI

Viscount Palmerston to Sir Charles Bagot, 13th January, 1831

Bagot MSS. Levens Hall, Westmorland.

F.O. 13 Jany 1831
recd. Jan. 16 1 PM

My dear Bagot,

is Capaccini still at The Hague, it occurs to me that if he is, he might if willing, be of great use to the P. of Orange.

When he was in London last I saw him and had some conversation with him on the affairs of Belgium; I inferred from what he said that he was friendly to the House of Orange, though fully alive to all the faults of conduct and character of the King, and convinced at that time that the P. of Orange had no possible chance of retaining Belgium without civil war. He however said that he thought the Belgians would have no objection to a Protestant prince and that a liberal Protestant for the very reason that he was a Protestant be perhaps more indulgent to the Catholic Church than some Catholic might be disposed to be.

I apprehend than in what he then said, he alluded to P. Leopold whom however I purposely abstained from mentioning to him. But Capaccini said he had gone to Brussels with the Prince out of personal regard. – Now it occurs to me that if Capaccini is aware of the great change which is said to have happened lately in opinion in Belgium as to the Prince, and if he knows that the general disposition of the nation is in his favor, he might perhaps be persuaded to use his influence with the priesthood to assist the Prince's cause.

You might therefore do well to have an interview with him as soon as possible and sound him on the subject. You will tell him that you had good reason to believe that the Orange party has gained great strength in Belgium, that *all* the Five Powers would gladly see the Prince elected as the speediest mode of arranging this very difficult and complicated affair. That hardly *any* other choice is in point of fact left; each court having some almost insurmountable objection to every other candidate and there is the greatest danger that if this question remains unsettled much longer it may involve Europe in war. That consequently in promoting this Orange arrangement he would be labouring in his own vocation as a peace maker and be rendering a most important service to Europe – The fact is that difficulties as to other choices have greatly assisted the Prince of Orange. The French party are to a certain degree defeated by the peremptorily [made] refusal of France to accept Belgium or to give Nemours; The Bavarian party find they cannot have P. Otto; Leopold is objected to by France; the Neapolitan Prince would not suit, and in the meantime during the time that these discussions are going on, distress increases, sufferings augment, and the Belgians wish for any settlement rather than none.

If you find Capaccini well disposed you might suggest to him that Brussels lies not much out of his way to England or Paris, to one of which places he said he meant to go when he should have finished his affairs at The Hague; and if he was willing to espouse the cause in any way you might shew him the Prince's letter or proclamation whichever it might be called which probably would satisfy him as to the Prince's views and intentions should he be called to govern Belgium.

Esterhazy who is with me while I am writing has promised to me to write a few lines to the same effect to Capaccini which he will inclose to you.

I see all the objections to the Prince on the score of his personal character and have great doubts whether the arrangement would last, but for the present it would preserve peace and that is a matter of much importance.

There will be many things to be settled besides, the commercial relations of the two countries, the exchanges of territory to be made for their mutual advantage and other matters of the same kind and the more discretion the King gives on those points to Falck and Zuylen the sooner and better the settlement can be made.

APPENDIX VII

Sir Charles Bagot to Viscount Palmerston, 18th January, 1831

Bagot MSS. Netherlands Correspondence Vol. IX, Levens Hall, Westmorland.

The Hague, Jany 18, 1831

My dear Palmerston,

The messenger Fricker arrived here at three o'clock in the afternoon of Saturday last, and delivered to me your letter of the 12th inst. On the following day at one o'clock Havilland arrived with your letter of the 13th. Half an hour after I received the latter, I had a private audience which I had previously requested of the King and which lasted one hour and three quarters. – This morning I have had a second, which lasted three hours and a quarter and in which I conscientiously believe that I have urged every argument which could be adduced in order to persuade The King to accede to the wishes of the Conference expressed in your two letters to me. – All has been in vain, and neither you, nor the Plenipotentiaries of the other Powers, nor the Prince of Orange must now calculate upon the possibility of the King being ever induced to renounce in the Prince's favor the sovereignty of Belgium. – Not only will he not do it contingently, and in the event of the crown being offered to him – but he will not do it at all.

It would be difficult and it would perhaps serve no end to attempt to report to you in detail the substance of two conversations which together lasted five hours, and in which there was I believe no point of the King's real or supposed position left untouched by him and by me.

In my first conversation with him I laid before him, in the strongest, and I believe the clearest terms the situation in which he stood – the absolute necessity of deciding instantly whether or not he would lend himself to the only project which could save the crown of Belgium to his family and the general wish of all the powers that he should do so. – I told him everything without the smallest reserve and I left him with a hope that I succeeded in making some impression upon him. – This hope was much strengthened by his sending M. Verstolk to me the next morning to say that he would see me again at 12 o'oclock today, and that he had been charged, in the meantime, to have some conversation with me as to the suggestions which I had thrown out to the King respecting compensation and the cession of Luxembourg to the Prince with a view to its incorporation with Belgium, – as well as to enquire of me whether I meant that my injunction of secrecy to The King should extend to M. Falck and M. de Zuylen.

In this conversation I explained in the fullest detail the wish and views of the Five Powers as regarded Luxembourg but told him that I had no distinct instruction as to the nature of the compensation in other respects which it might be in contemplation to propose for the advantage of both Parties. – It is true that your letter adverted to pecuniary compensation,but seeing as I thought, a disposition to yield Luxembourg I thought it prudent not to chill this disposition by crushing all hopes of territorial compensation, if such an idea should be in the King's mind. That it was in his mind I soon collected from M. Verstolk who threw out that for Luxembourg the King might be willing to take the Province of Limbourg, certain parts of Liège – containing the coal and iron mines nearest to his frontiers and whose chief debouché for their product was Holland – and the Forts of Lillo and Liefkenshoek on the Scheldt which originally belonged to Holland, and had been ceded to Austria by the Treaty of Fontainbleau – All this of course I took ad referendum and without observation – but I own that I flattered myself with having brought matters into a train of negotiation which might lead to some good issue as regarded Belgium and the Prince's cause.

My audience of today has dissipated *every* hope that I had. – You will scarcely credit it, but the King is at this hour as far from having divested himself of the idea of its being possible that he should yet retain, in his own person, the sovereignty of Belgium, as he was on the 25th of last August. He will write to the Prince (I believe to night) to tell him that he is at liberty to do what he pleases, or what he can, in Belgium, as his Lieutenant Général Provisoire – but beyond this nothing will induce him to go. He never adverted again either to Luxembourg or compensation, or any of the points respecting which he had charged M. de Verstolk to talk with me excepting indeed as regarded his communicating upon the subject of Prince of Orange's matters with M. Falck and M. de Zuylen which I told him that he was certainly at liberty to do, and that indeed the Plenipotentiaries anxiously desired that they should be furnished by His Majesty with the fullest instructions, discretion and powers upon every point which could come into discussion.

Whether this change of the King's mind from what it appeared to be to me on Sunday was to be attributed to the arrival of the Queen with whom he may have conversed – or to the information which he had perhaps (though I do not think he had) received at the time of the proceeding at Ghent, and to a printed address in his favour from the Town of Liège which lay upon his table and which he begged me to read, I do not know – but you may rely upon it that his mind is made up to take all chances rather than listen to any voluntary renunciation of his Belgian Sovereignty. – He was in several parts of the conversation affected deeply and to tears. The most remarkable observation which fell from him were that perhaps the better course would be to abdicate entirely in his son's favor – as in that way he might still preserve unimparted to his family the rights of his House – and at one moment he said "Why not throw myself into the arms of France? – she might have pity for me. Successful examples of such a proceeding are not wanting even in modern history. The Elector of Bavaria and the Grand Duke of Wurtemburg acted so in regard to Napoleon, followed him in his career and preserved their thrones." – Can it be that this is the secret hope and purpose of the King? – God knows – nothing is clear to me but my entire failure in the business which the Conference have confided to me. The King is inmoveable.

You will have learnt soon after you despatched your letter that Capaccini was in London. I return enclosed Prince Esterhazy's letter to him.

This is a very hurried – perhaps a very confused letter – but I am writing against time – as I wish to despatch one of the Messengers by Calais with duplicates of what I send by the post. – I shall most probably redespatch the other Messenger to morrow night.

Charles Bagot

APPENDIX VIII

Viscount Granville to Viscount Palmerston, 25th March 1831
Broadlands Archives

Private Paris, March 25, 1831

My dear Palmerston,

It was quite amusing to me this day to hear Sebastiani taking credit to himself for having brought to maturity his Leopold project; you see, said he, with what steadiness I pursue any plan which I have taken in hand; and dites à Lord Pal-mer-ston that I have the greater gratification in proposing it at this moment, because it is calculated to fortify the administration in England in the crisis in public affairs caused by the division of a majority of one on the great question of Reform.

There are rumours that an explosion is about to take place at Antwerp in favour of the Prince of Orange; and the extreme anxiety which the French Government now shows to hasten the election of a Sovereign by the Congress, and the perfect confidence they feel that Prince Leopold will be elected by that Assembly, indicate pretty plainly that they apprehend an Orangist movement. – The proposal of Leopold will not be popular here and the Ministers are expecting to be furiously attacked by the Press and in the Chamber for their adoption of it; but I really believe they think it will facilitate the preservation of peace. – All the Belgian provinces bordering upon France will resist the attempt of Antwerp and Ghent to place the Prince of Orange on the throne; the French Government, if it does not openly interfere in favour of the neighbouring provinces, will connive at assistance being given to them, because the elevation of the Prince of Orange would be under present circumstances the defeat of the French party; and the result will be a general war, or a partition of Belgium, by which France would acquire an extension of territory in Flanders, Holland would gain in Brabant, and Prussia obtain a slice on the side of Liège and Luxembourg. – I apprehend, tho', the whole negociation will be left very much in the hands of Talleyrand. Baron Louis in speaking to me last night on the Luxembourg question said, we had better leave it to be settled by the Conference in London.

Bresson is now at Paris – he explained to me of the embarrassing position, in which he had been placed at Brussels; the instructions he observed which Lord Ponsonby received from his Government, accorded entirely with those, which in his character of Commissary he received from the Conference, whilst those which were sent to me from Paris, directly contradicted those which came from London. – I had to choose which I would obey; and I thought it my duty to execute the orders which came immediately from my own Government.

I am much pleased with your instructions both to Lord Cowley and to Lord Heytesbury; but if it be true that the Austrians are taking possession of Bologna, they are treating the Government most scurvily, and acting on a system of policy which may end ill for them. Why should Prussia, why should Russia assist them, when rejecting the moderate propositions of France, and heedless of the advice of England, they go on obstinately lending their military aid to keep in abject subjection the whole Italian Peninsula.

I believe that Austria, Prussia and France, all concur in the opinion stated in your Despatch to Lord Heytesbury regarding the incorporation of Poland with the Russian dominions.

You will see Walewski in England, he showed me a despatch from Nesselrode to Grand Duke Constantine, which had been left at Warsaw, in which speaking of King Leopold being named King of Greece a year ago, he says, "Loin d'avoir contracté avec [sic] affection pour l'Angleterre, il désire la quitter parce que ses relations avec le Roi, le Ministère, *et les Pays Bas* en général sont pénibles et fausses." Perhaps as the two former have been changed, he may not be at present anxious to be seated on a foreign throne.

Yours ever sincerely,
Granville

APPENDIX IX

Viscount Palmerston to Mr. G. W. Chad, 3rd May, 1831

Broadlands Archives

Private Cambridge, May 3. 1831

My dear Sir,

I avail myself of the departure of Mr Magenis to write to you a few lines, though in the midst of the occupations consequent upon a contested election, and although I am unable to give, what I have to say the shape of an official dispatch, without detaining Magenis some days longer.

The Government are led to think that it may not be impossible that the Emperor of Russia should make some proposition to Prussia to afford him military aid against the Poles, and it is needless to say, that we feel anxious to dissuade the Prussian Government from giving such assistance if it should be asked to do so. – We know from Vienna that Count Bernstorff would be disposed to comply with it, and that Prince Metternich leans to the same opinion, but that the majority of the Austrian Cabinet decided upon remaining neutral. We wish you to endeavour to ascertain whether any such proposal has been hinted to Prussia by the Russian Government, and if we should have reason to think it has been or may be made, you will represent in the same friendly manner to the Prussian Government the tendency which such an interference on their part, beyond their own territories, in Poland...[12] to embroil them with France or to produce commotions in the Rhenish provinces; and you will say how much concerned we should be to see a rupture between France and Prussia upon such grounds, because with all our desire to assist Prussia and our

[12] One word illegible.

deep sense of the importance of maintaining her in her present territory, we fear the public opinion in England would render it absolutely impossible to go to war with France, in order to leave Prussia free to join Russia in crushing the Poles. I mentioned this view sometime ago to Baron Bulow, and begged him already to explain to his Government that if they were to involve themselves in war in consequence of their assisting the Poles [13], no English Government would be able to afford them any assistance, and there is scarcely anything which would be more painful to the present administration than to see Prussia involved in difficulties with France and yet be unable to give her any support. Baron Bulow said he would report what I said to his Government not in the shape of a communication to them, but as a statement made by me in conversation to him; and if you should have an opportunity of making these remarks to Count Bernstorff, it is desirable you should do so. I had a similar conversation with Prince Esterhazy before I received Lord Cowley's letter informing me that Prussia had sounded Austria upon the question of interference, and that Austria had declined, I represented to Prince Esterhazy in the same manner as to Baron Bulow the embarrassments which might arise to Europe from their interference by arms in the Polish war and how desirable it was for them having just got out of the Italian question not to invite new difficulties for themselves about Poland. I also suggested to him for the consideration of his Government that if the Emperor of Russia should ask for military aid, the Austrian Government might very properly avail itself of the opening which would be thus afforded to offer mediation instead of arms; and I wish you would hint the same thing to the Prussian Cabinet.

Let it not be supposed that the English Government is hostile to Russia because we wish to prevent the extremities to which the war in Poland has a tendency to go; we think we should be doing great service to Russia, could we bring about an accomodation which would restore things to the state in which they were previous to the revolution, by a treaty or peace negociated by the intervention of friends. When a whole nation rises in arms and fights a campaign the matter ceases to be an insurrection which is to end by submission and punishment, it becomes a war which can only be terminated by treaty and concessions, and the dignity of the Emperor would not

[13] Cf Chapter III, above, p. 140.

allow him to negociate with those he considers as rebels, the intervention of friends might save his honour. That mediation however cannot be afforded until some application for assistance or intervention is made by the Emperor, or at least till the experience of a longer period shall have shown that Russia is unable without enormous sacrifices to reconquer the Poles. An accomodation upon the footing that on the one hand the Emperor should forgive men and restore institions, and that on the other the Poles should resume allegiance and return to obedience would give to Russia all she could legitimately obtain by the present success because she could not consistently with the Treaty of Vienna place Poland in a secure state of dependance upon an incorporation with Russia than to make it attached to Russia by its *constitution* with a *separate administration.* But to re-establish this order of things by force of arms upon the principles laid down by Russia, which is, that the Poles must submit unconditionally and trust to the mercy of their conqueror, a thing they never can be expected to do till driven to the verge of extermination. To accomplish this, I say, by arms, will cost Russia an enormous sacrifice of blood and treasury as she must remember that in this encounter like in an aimaible suit in Chancery the costs on both sides really fall on her: because if she is to possess Poland in the end, every waste of Polish resources is a loss ultimately to Russia.

The Prussian Government feel differently upon this question from the Austrian – Austria would be glad to see an independant Poland, and would probably sacrifice Galicia to obtain it, however different her language may be at present for the purpose of keeping fair with Russia. But Russia surrounds her frontier upon so many points, that it would be a great relief to her to get a new neighbour, at least upon one side.

Prussia, on the other hand looks I conceive to Russia, more as a support than as a source of danger; she never contemplates the possibility of being able to cope with Russia, and is prepared to follow the Russian politics, . . .[14] upon the Emperor as a sure ally in the always probable event of a quarrel with France. Prussia therefore would think it a loss to have interposed between her and Russia a constitutional Power, upon which she could not equally count, and whose sympathies would be more likely to be with

[14] One word illegible.

France than with Prussia. The Prussian Government would also fear that if Poland were to become independant the Duchy of Posen would sooner or later be united to it, and then old Prussia would be cut off from the rest of the Prussian dominions. But all their considerations ought rather to incline the Prussian Government, if well advised, to desire the contest brought to an end by accommodation, upon the footing of restoring things to the condition in which they were before the revolution, because they ought to feel that every month the war continues, the chances of an issue adverse to Russia encrease – besides the military difficulties in Poland which augment with time, because the Poles get more organised, better armed, more deeply and universally pledged, while the Russians suffer from wear and tear at a distance from their resources; besides these things must be considered the chances of further insurrection in Russian Poland, and even in Russia itself...[15] the possibility of a change of policy in France leading the French Government to some positive demonstration moral or physical in favour of the Poles. Still we fully acknowledge that in the present state of this business it would be unbecoming for other Governments to offer to interfere, but everybody knows, how a few words incidentally thrown in, in casual conversation may convey an opinion which could not be stated formally without offence.

Sebastiani throws upon Guilleminot the blame of the intrigue at at Constantinople[16] against Russia, but it seems doubtful whether dates will beat him out.

My dear Sir.
Yours faithfully,
Palmerston

[15] One word illegible.

[16] The *démarche* of the French Ambassador in Constantinople, Guilleminot, is described in the introduction to Chapter IV, see above, p. 143.

APPENDIX X

The interview of Viscount Palmerston and the Polish agent in London, Count Alexander Colonna-Walewski on 15th July, 1831

Extract from a letter of Walewski to the Polish Government, dated July 18, 1831. (MSS of the Bibliothèque Polonaise in Paris, Diplomatic Documents vol. XI, Cat. No. 355).

Je commençai à lui représenter de la manière la plus vive combien nous avait été nuisible la manière dont la Prusse manquait à la neutralité à laquelle elle s'était engagée à notre égard. Je lui assurai que sans partialité de cette puissance depuis longtemps le théâtre de la guerre ne serait plus Pologne et qu'enfin, si une guerre générale venait à avoir lieu à cause de la Pologne c'est tout à fait à la Prusse et à l'Autriche qu'on devrait s'en prendre pour n'avoir pas tenu et peut-être aussi à l'Angleterre pour avoir laissé faire. J'ai trouvé Lord Palmerston assez bien préparé et disposé à ce sujet; il est entré dans toutes mes raisons et a ajouté: "Dans ce moment nous avons une bonne entrée en matière, car différentes réclamations nous sont parvenues sur la conduite de la Prusse envers nos propres sujets, nous allons nous servir de ce prétexte pour faire des représentations qui j'espère seroient fructueuses. Nous en avons faites de très positives à l'égard du Gal Dwernicki mais jusqu'au présent pas de réponse." Je lui parlais alors de la conclusion des affaires de Belgique et de l'influence que nous y avons exercée. Je lui rappelais la conversation que j'avais eue avec lui à ce sujet et lui demandai si le moment n'était pas venu enfin d'intervenir, de médiatiser dans les affaires de la Pologne. J'ajoutai enfin qu'entre mille autres raisons, il y a une qui me donnait l'espoir qu'enfin l'Angleterre parlerait car c'était un seul moyen d'empêcher une guerre Européenne. Vous connaissez, lui dis-je, la position de la France, vous devez savoir tout aussi bien que moi qu'aujourd'hui à la question polonaise est attachée l'existence du ministère actuel. Le ministère renversé, vous ne pouvez douter d'une guerre générale. Dans le cas contraire, si la France est obligée d'agir seule à la Pologne, cette intervention inspirera des craintes à l'Autriche et à la Prusse et encore guerre générale s'en suivra, mais si au lieu de cela, de concert avec le cabinet français, vous élevez la voix d'une manière ferme et décidée, je ne doute pas que la Russie ne cède à vos démarches, quelques dures qu'elles puissent lui paraître; si non, au moins l'Autriche et la Prusse ne se mêleront de rien n'ayant plus rien à craindre de la propagande française, puisque vous qui seriez à la tête de cette intervention, soit qu'elle devienne armée, soit qu'elle soit pacifique.

Palmerston me répondit: "C'est la raison principale qui nous engagerait à intervenir dans vos affaires, nous y avons pensé; nous y pensons encore; mais nous croyons que le moment n'est pas tout à fait propre. S'il ne s'agissoit que d'envoyer un courrier à St. Pétersbourg, comme l'a fait la France dernièrement, pour offrir à la Russie nos bons offices, nous le ferions demain, mais qu'en résulterait-il? nous aurions la même réponse que a eu [sic] la France, et nous ne voulons nous exposer à avoir une pareille réponse. Quand nous pourrions dire à la Russie (et mon opinion personnelle est que cela arrivera incessamment): Vous avez déployé en Pologne toutes vos forces; elles ont été infructueuses pour arrêter insurrection. Vos troupes, malgré tous les renforts que vous leur avez envoyés, ont été forcées d'évacuer le pays; des insurrections dans toutes les provinces Russo-Polonaises et même dans l'intérieur de votre empire rendent tous les jours votre position plus critique; enfin il y a incompatibilité et vous n'avez plus la force nécessaire pour faire rentrer dans l'ordre ceux de vos sujets qui se sont révoltés. Une seconde campagne ne peut pas avoir lieu. L'Europe ne peut pas voir se déchirer ainsi deux nations à l'existence desquelles elle est intéressée et nous sommes obligés d'intervenir. Quand nous pourrons tenir ce langage à la Russie le moment d'intervenir sera venu, mais aujourd' hui si nous le fassions la Russie nous dira: nos ressources sont loin d'être épuisées, quelques semaines encore et tout sera fini, de nouveaux renforts viennent d'arriver au nouveau chef et dans peu nos troupes seront à Varsovie. Je crois donc, a ajouté Lord Palmerston, que pour agir avec droit, il faut que nous attendions encore quelque tems. Je lui représentai alors combien tout délai était déplorable pour l'humanité entière, etc. etc. Mais tous mes raisonnements ont été inutiles, il est revenu au mêmes arguments. Enfin je tâchais de pressentir si le Cabinet Anglais laisserait agir librement la France.

J'ai cru apercevoir qu'oui, quoique Lord Palmerston m'ait dit que cela dépendrait pourtant du genre d'intervention que voudrait entreprendre le Cabinet Français. En me quittant il me renouvela la promesse de s'occuper sans relâche de la Prusse et me répéta l'assurance de tout l'intérêt personnel qu'il prenait à la cause polonaise. "Je vous assure", m'a-t-il dit, "que je n'ai pas besoin d'être poussé."

Quand je rapportai cette conversation au Prince de Talleyrand, en partie du moins, car je ne lui dis rien de ce qui avait rapport à laisser agir la France seule, il frappa sur la table et me dit: "Les Anglais sont toujours les mêmes, toujours la même politique mais cela ne fait rien. Il faut qu'ils nous répondent et je n'aurais pas de refus."

APPENDIX XI

Lord Holland to Viscount Granville, 19th July, 1831

Granville MSS at the Public Record Office, London, Ref. No. P.R.O. 30/29/9.

My dear Granville,

I cannot agree with you on the impropriety and still less on the hazard of some joint representations about the Poles to Russia. You all seem to me to underrate the authority which France and England when acting jointly can speak to all other Powers and to Russia in particular and on this side of the water do not think as strongly as I do that we are bound in honour and in policy to do all that we honestly can to strengthen Casimir Périer and to counteract the loss of popularity he may have sustained by the election of Leopold and they are not aware of the assistance which anything like co-operation about Poland would offer him in conciliating publick opinion. France must say or do something on that subject whether we do so or not, – but if we will act conjointly with her, she will no doubt regulate her language and actions in a very great measure by or wishes. Is that not a great advantage? – and is it not more *likely to preserve peace* than leaving her to act alone with no other direction but that which the current of popular breath may give her? And will not concurrence with her on this point make it easier for her to yield to, and give us greater authority in making any representation which the state of her relations or hostilities...[17] may require? It does not appear to me that an offer of mediation or a friendly admonition to Russia implies war in case of refusal, though it may inquestionably suggest to the mind of the Emperor the wholesome hint of consequences which war, if ultimately provoked by his conduct in Europe, would inevitably produce. We

[17] Two words illegible.

offered to mediate between Spain and her colonies. – We apprized her that if she declined, recognition of those colonies by England might be the consequence – Spain refused – recognition not war was the only result. Why would not France and England jointly follow the same course about Poland and Russia – I believe, if they did, the permanent independence of part of Poland would be the consequence and I further believe that Austria and perhaps Prussia would not in her heart be grieved at the result – I am confident they would not advise or concur in a war against France acting with England and I am equally confident that without their assistance Russia neither would nor could undertake it. When it is known, as known it will be, that France shows at St. Petersburgh more interest about the Poles than we do, the publick opinion in favour of our joining their councils will be so strong that those who are most averse to it now, will be quite unwilling to resist it – but then France not England will have all the grace of mediation and any power between Germany and Russia which may be revived in consequence of such interference will form more emphatically part of the French rather than the English connection in Europe.

I have been bound to bore you with a long argument but I am growing very eager on the subject – nor have I quite given up the reciprocal search – Talleyrand says we ought to have it. – Carlisle has gout and does not look well.

Yours ever
Vassall Holland

APPENDIX XII

Lord Holland, memorandum for the Cabinet meeting on 25th August, 1831

Broadlands Archives, File 3. Cabinet Ministers.

It is I think clear that the King cannot receive the Deputies and the letter which they wish to present to him without a virtual acknowledgement of the provisional Government of Poland, I cannot say that he therefore should be advised not to receive them because I am not prepared to say that in the present circumstances he ought not to recognize that Government, and place this country in diplomatick or at least commercial relations with Poland, and on the contrary I think that this question [is] well deserving the early consideration of the Cabinet. – Till, however, we have determined on that step we cannot advise to the King to receive the letter from the hands of these deputies to whom it has been entrusted. How far there is a difference in receiving it when transmitted through other hands or how far the bare receipt of a letter signed by a person in a particular capacity can be construed as an acknowledgment of his title to that character is a nice question which must depend upon diplomatick usage and construction and on which I do not venture to give an opinion. – In common sense, however, there seems to me to be this obvious distinction. By receiving a person avowedly deputed by another and still more by receiving from that deputation so avowed a formal letter or message I imply that I recognize the authority of the person who had sent it, but by receiving a letter through an unknown or unauthorized channel I cannot express any opinion of the character or capacity in which it is written for I am not reasonably supposed to know either the contents or the writer till I have read it.

I therefore should after declining to receive the deputies feel no scruple in receiving the letter.

APPENDIX XIII

A.R. Falck and Baron van Zuylen to Baron Verstolk van Soelen, 20th September 1831

Extract from a dispatch of the Dutch plenipotentiaries at the London Conference on Belgian affairs, Ambassador A. R. Falck and Baron H. van Zuylen van Nyevelt to the Netherlands Minister for Foreign Affairs, Baron J. G. Verstolk van Soelen. (Algemeen Rijksarchief, The Hague, Ref. No. 1696 (B.Z., Conferentie Londen).

No. 81 Londres, 20 septembre 1831

La nouvelle de la prise de Varsovie donne matière à de chaleureux articles dans presque tous les journaux, destinés a manifester le plus vif intérêt pour la cause des Polonais et à inculper plus ou moins les Ministres de l'avoir abandonnée. En général, le succès de la Russie a fait une impression d'autant plus vive sur eux que jusqu'il y a peu de tems, ils étaient persuadés que la Pologne était perdue pour Elle. Lord Grey et Lord Palmerston surtout ne cachaient pas leur sentiment à cet égard. Bien que personne ne se dissimule que cet événement doive nécessairement influer sur la négociation qui nous concerne, il est évident que l'on veut ne pas se l'avouer. Aussi nous apprenons qu'une de premières démarches des Plénipotentiaires Russes, après la réception de la nouvelle, a été de témoigner à Lord Palmerston et à M. de Talleyrand, que cet événement ne changerait en rien les dispositions de la Russie dans la question Belge.

Cette déclaration verbale est peut-être justifiée par les ménagemens que l'Angleterre a mis dans sa protection en faveur des Polonais, ménagemens qui ont vraisemblablement retenu la France à intéresser pour Eux d'une manière plus énergique: sous ce rapport, elle serait dans les formes de diplomatie, mais il nous reste à désirer qu'elle ne soit pas suivie à la lettre. Aussi, depuis que par hasard nous en sommes informés, nous nous efforçons de rappeler à nos amis que pendant neuf mois les affaires de la Pologne nous ont été représentées par Eux comme la cause directe et principale des empêchemens de l'Europe à nous favoriser; qu'il était donc naturel qu'un événement qui remettait la Russie, la Prusse et l'Autriche en état de faire agir des armées paralysées jusqu'ici, nous donnait l'espoir de voir ces puissances, non pas précisément revenir sur ce qui était déjà accompli, mais soutenir nos justes prétentions dans les futurs arrangemens définitifs.

Il nous semblerait effectivement fort dur, après avoir été victime de leur faiblesse, de la devenir maintenant de leur générosité.

APPENDIX XIV

Viscount Palmerston to Mr. G.W. Chad, 30th December, 1831

Broadlands Archives

Private — Foreign Office, December 30th 1831

My dear Sir,

I never was much more surprized than I was at the receipt of your last dispatches, informing me that upon the first rumour of the refusal of the Emperor Nicholas to ratify the Treaty, M. Ancillon was preparing to eat up all his words and to back out of all his engagements about Belgium. – He certainly has a difficult task to perform in this respect, and would deserve some applause if he could acquit himself with decent success, but what he has undertaken is beyond the powers of man, for no human ingenuity could invent a plausible pretence upon which Russia could now refuse to ratify the Belgian Treaty.

I know not who furnished M. Ancillon with his new dictum about treaties, nor whether he gave it to some, from whom he have [sic] heard it in this Town within the last four or five days, or whether they suggested it to him, but the proposition is untenable, but who may be its author.

The first sentiment it produced was extreme indignation, the next a feeling not so nearly balanced with respect. You are authorized to communicate to M. Ancillon the whole of my dispatch, and it is even desirable that you should; and if he wishes it, you may even leave him a copy of it. You will tell him that Parliament meets on the 17th of January, that it is absolutely necessary we should be able on that day to inform the two Houses, how we stand with our Allies in respect to the ratification of the Treaty, the signature of which was announced in the speech from the Throne at the opening of the session; England cannot refuse to ratify. There is no pleading for her that instructions have been contravened or powers overstepped, and it is only upon those grounds that an honourable Government can refuse to ratify a treaty. –

Notwithstanding the little breeze we have had about the Fortresses I have no doubt whatever that France will ratify also – England, France and Belgium, then will have to carry into execution the stipulations of the Treaty –

What will Prussia do? will she oppose them and range herself under the banners of Holland, to prevent the execution of an arrangement which she has pronounced to be equitable and just and which she has even promised to ratify – this is not very probable. Will she stand by then an inactive spectator and incur all the odium of broken faith, without any advantage to the Party, on whose behalf she would have violated her engagements? These are questions which M. Ancillon ought seriously to consider because they will very shortly force themselves upon him. –

I see nothing that can prevent a war except the ratification of the Treaty. – If Prussia refuses, one or other of two things will be demonstrated; either that in spite of her pacific professions Prussia is panting for war; or else that she is not an independent agent, and is dragged by Russia at her chariot wheels – You must insist upon having an immediate answer, Yes or No, will Prussia ratify or not, and it should be given in a Prussian Note in order that it may be produced, if it should become necessary to do so, if the Papers are laid before Parliament; which upon all occasions of unsuccessful negociations [they] invariably are.

As the time is barely long enough to admit of a courrier going and returning, I have settled with Talleyrand, that in a matter of common interest to France and England we shall have the use of the French telegraph from Strasburg through Paris to Calais. Talleyrand writes to Bresson to desire him to communicate with you and tell you how it is to be done; but I imagine that you will send to Strasburgh a short message, such as a telegraph can readily work such as "Prussia ratifies immediately", your dispatch will come in the usual way, but the telegraph message will probably reach London some days before your courrier. –

You will not keep your courrier till you can say that Prussia has her ratification made out, but get a short Note stating in writing her intention, and send off your messenger as soon as you get it. –

We are preparing an answer to the long Note of the Dutch

plenipotentiaries and I have no doubts of our being able successfully to refute all their assertions.

Tell M. Ancillon that we are beginning to regret that we did not take a more decided part in Polish affairs before the capture of Warsaw, we might have done some good in Poland, and have had less trouble with this Treaty. –

My dear Sir,
Yours sincerely,
Palmerston

THE NETHERLANDS 1815-1839

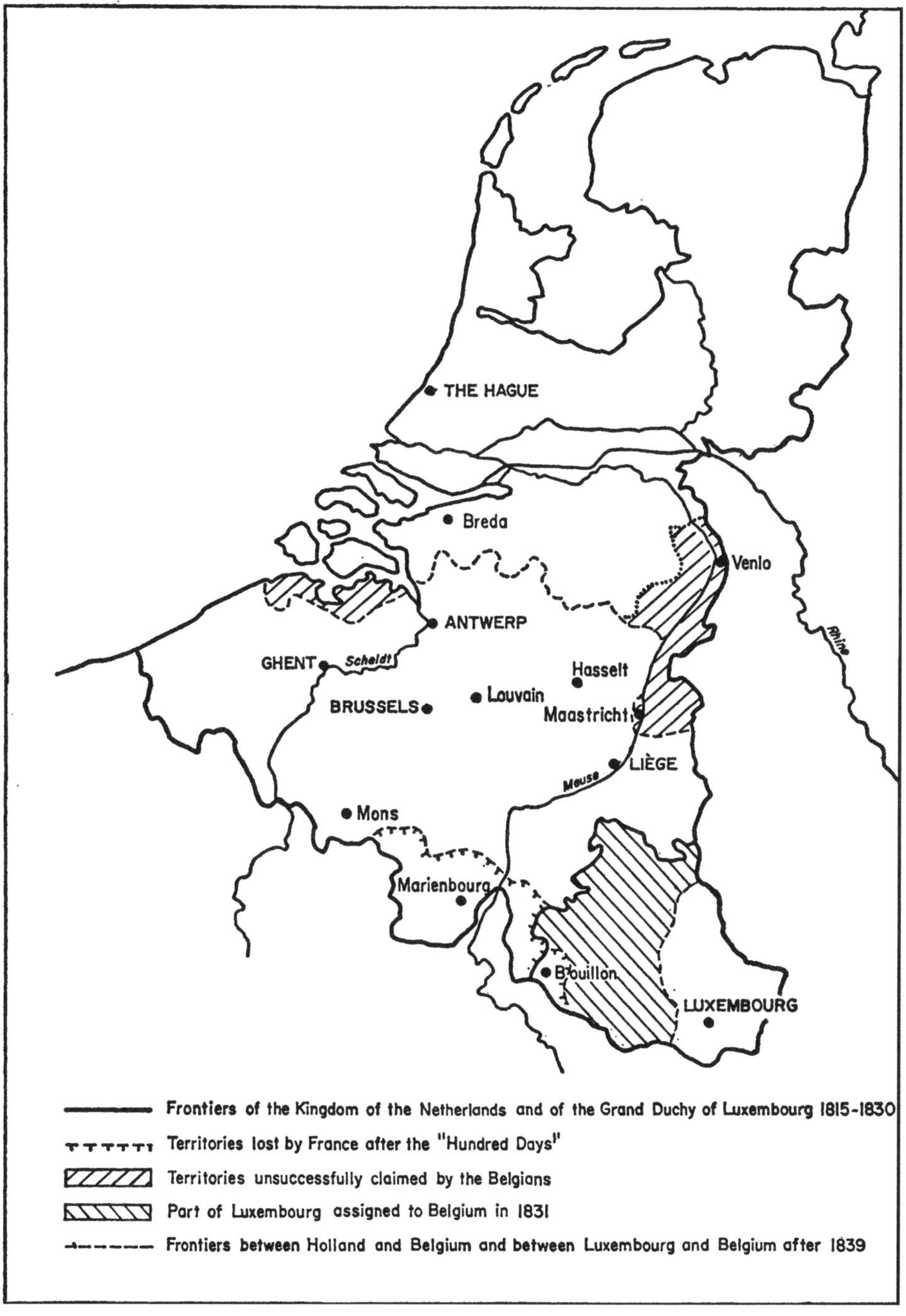

POLAND 1772-1830

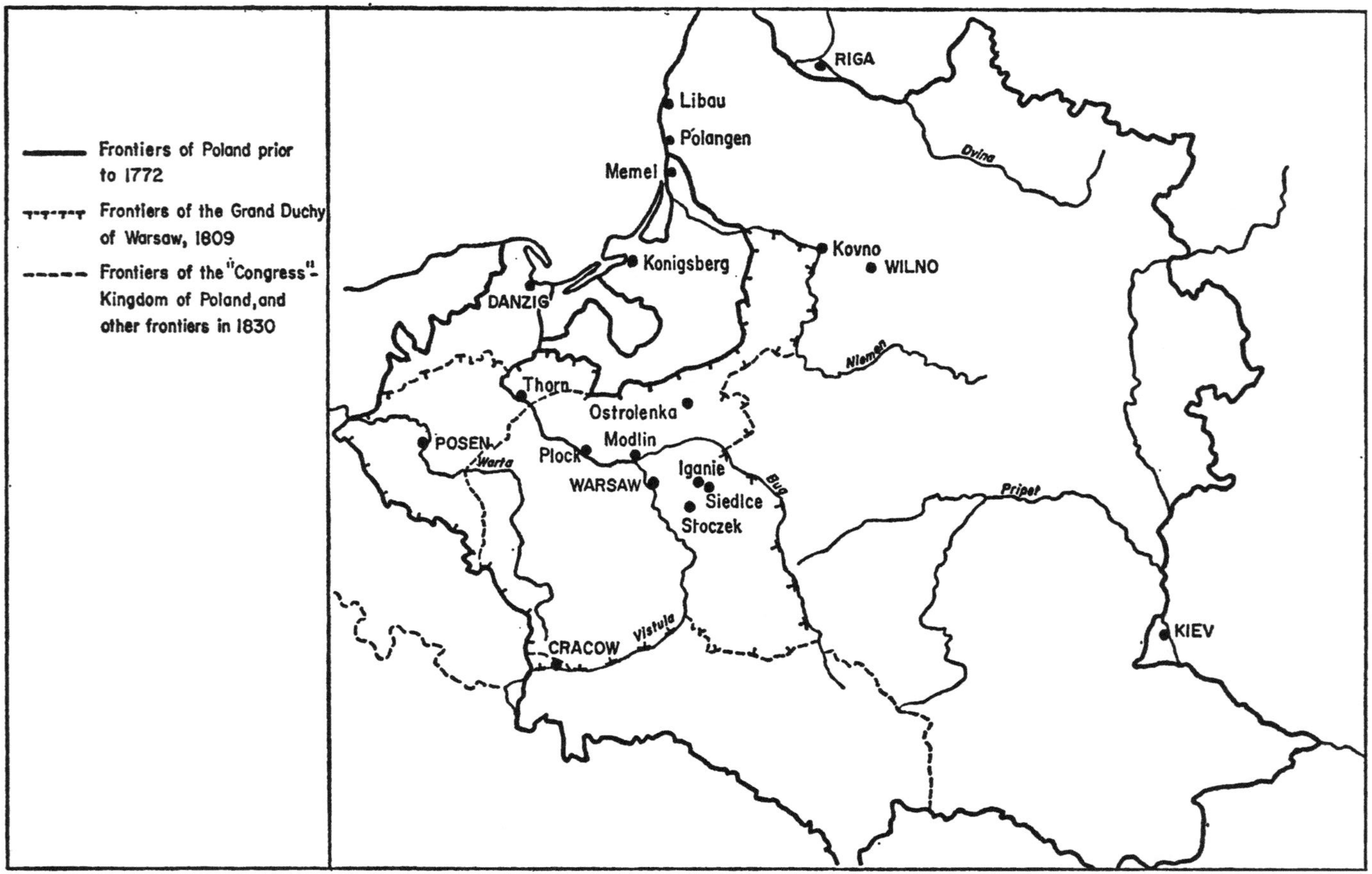

INDEX

Figures in brackets indicate that a person is mentioned by title only

Bei Fragen zur Produktsicherheit wenden Sie sich bitte an:
If you have any questions regarding product safety,
please contact:

Walter de Gruyter GmbH
Genthiner Straße 13
10785 Berlin
productsafety@degruyterbrill.com